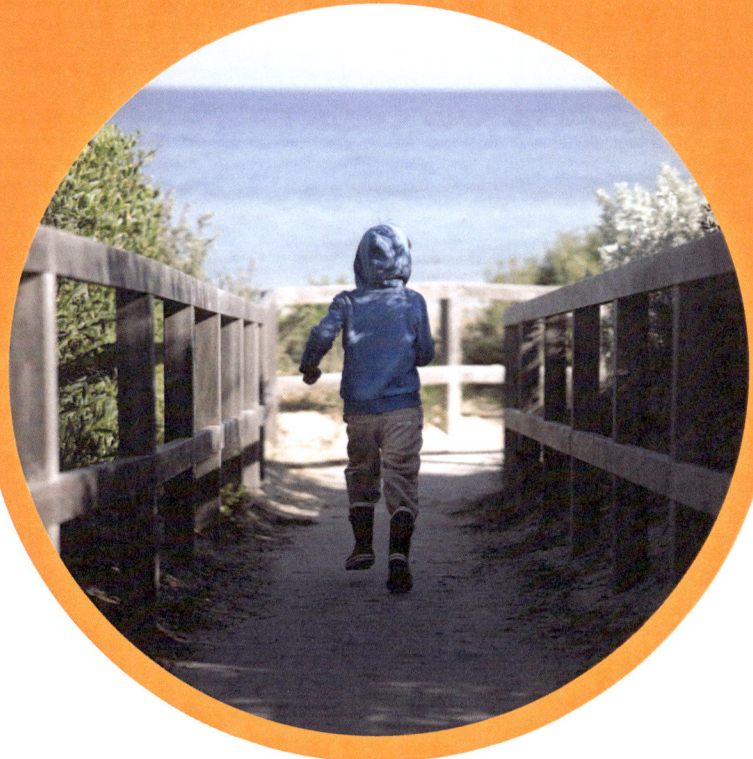

ECOLOGY FOR HEALTH

Design Guidance for Fostering Human Health and Biodiversity in Cities

SEPTEMBER 2023

FUNDED BY
THE ROBERT WOOD JOHNSON FOUNDATION

PRIMARY AUTHORS
Joe Burg
Jennifer Symonds
Vanessa Lee
Bronwen Stanford
Brandon Herman
Stephanie Panlasigui
Karen Verpeet

DESIGN AND PRODUCTION
Ruth Askevold
Joe Burg
Jennifer Symonds
Brandon Herman

URBAN DESIGN GRAPHICS
Vanessa Lee
Joe Burg
Brandon Herman

SFEI
PUBLICATION #1130

I0092860

SFEI
San Francisco Estuary Institute

PREPARED BY San Francisco Estuary Institute

IN COOPERATION WITH AND FUNDED BY THE
THE ROBERT WOOD JOHNSON FOUNDATION

SUGGESTED CITATION

San Francisco Estuary Institute. 2023 Ecology for Health: Design Guidance for Fostering Human Health and Biodiversity in Cities. Funded by the Robert Wood Johnson Foundation. SFEI Publication #1130, San Francisco Estuary Institute, Richmond, CA.

Version 1.2 (09/05/23)

REPORT AVAILABILITY

Report is available online at sfei.org/projects/ecology-health

IMAGE PERMISSION

Permissions rights for images used in this publication have been specifically acquired for one-time use in this publication only. Further use or reproduction is prohibited without express written permission from the individual or institution credited. For permissions and reproductions inquiries, please contact the responsible source directly.

COVER and TITLE PAGE IMAGE CREDITS

(front cover, counterclockwise to center) Park playground (Honey Yanibel, Unsplash); Two-tailed swallowtail *(Papilio multicaudata)* in Mexico City, Mexico (Don Fabia, Unsplash); Lincoln Park South Pond in Chicago (Ranjit Souri, Unsplash); Children playing soccer in grassy field, Jakarta (Robert Collins, Unsplash)

(title page) Beach trail (Mike Stevens, Unsplash)

(back cover) Design illustration (Vanessa Lee, SFEI)

CONTENTS

❶ San Antonio River Walk in San Antonio, TX (Hannah Bennett, Unsplash)
❷ Wildlife-friendly backyard garden (Carol Norquist, CC BY 4.0)
❸ Bangladesh National Botanical Garden in Dhaka, Bangladesh (Jab Shwadhin, Unsplash)

▲ Great blue heron, Vancouver in background (James Wheeler, Unsplash)

ACKNOWLEDGEMENTS

Support for this project was provided by a grant from the Robert Wood Johnson Foundation. Their funding enabled us to synthesize the fields of ecology and human health in a peer-reviewed journal article along with this document's design guidance recommendations. We are grateful to the Robert Wood Johnson Foundation staff who contributed to the project by providing guidance, advice, and collaboration throughout the project, including Paul Tarini and Michael Painter.

This project benefited from the range of expertises offered by our interdisciplinary Technical Advisory Committee (TAC). TAC members included Leslie Aguayo (Greenlining Institute), Myla Aronson (Rutgers University), Peter Bosslemann (UC Berkeley), Laura Crescimano (SITELAB Urban Studio), Iryna Dronova (UC Berkeley), Christopher Guillard (CMG Landscape Architecture), Howard Frumkin (University of Washington), Veniece Jennings (Environmental Leadership Program), Michelle Kondo (US Forest Service), Ming Kuo (University of Illinois at Urbana-Champaign), Daphne Miller (UC San Francisco, UC Berkeley), David Nowak (US Forest Service), Michael Painter (Robert Wood Johnson Foundation), and Alessandro Rigolon (University of Utah).

We acknowledge the Google Ecology Program, the City of East Palo Alto, and Canopy for their foundational participation in the early collaborations related to this design guidance document.

Finally, we thank the SFEI staff members who contributed to this project along the way, including Micaela Bazo, Matt Benjamin, Letitia Grenier, Robin Grossinger, and Erica Spotswood. §

01
INTRODUCTION

OVERVIEW

Greenspaces provide crucial nature contact for urban residents. When we have greater access and exposure to nature in the places where we live, work, learn, and play, we tend to experience better human health outcomes. Urban parks, trees, and vegetation encourage physical activity, reduce anxiety and depression, support social cohesion by providing gathering spaces, and are associated with reduced mortality and improved overall health.

While traditionally biodiversity conservation has focused on large open spaces, cities can also play a key role in supporting biodiversity. Many of the world's major cities developed in biodiversity hotspots due to historical settlement patterns dependent on natural resources. Thus cities contain vital remnant habitat as well as globally important native and endangered species that rely on urban greenspaces.

As urbanization increases, cities around the world are developing and implementing plans to better integrate nature within urban settings. Many of these plans emphasize the importance of urban greening in providing multiple, substantial benefits such as biodiversity conservation, stormwater management, human health and well-being improvements, climate resilience, and more. However not all greenspaces are created equal in their biodiversity support and human health provision.

The goal of this document is to provide science-based **guidance for designing urban spaces that foster both human health and urban biodiversity**. Anyone making decisions about land use and urban design in cities across the world can benefit from the recommendations in this document (including community organizations, local non-profits, local leaders and policy makers, city planners, urban designers, landscape architects, engineers, gardeners / horticulturists / arborists, residents, and landowners). However, the majority of the document is specifically aimed at supporting designers and planners who work at the **planning, site, and detailed design scales** such as landscape architects, civil engineers, and urban designers. As noted in more detail in the limitations section below, this document synthesizes global research and design strategies while strongly informed by our experience as scientists and designers in California's San Francisco Bay Area.

◀ (*Left*) Pedestrians crossing at an intersection in Shibuya, Japan (Ryoji Iwata, Unsplash)
(*Right*) Walkway by river in Rome, Italy (Mark Harpur, Unsplash)

We created this document to highlight opportunities for urban design and planning to create spaces that benefit both native wildlife and human health. A large and growing body of literature documents the ability of cities to support native plants and animals. A parallel body of literature describes the human health benefits associated with access to biodiverse greenspace. Many synergies exist between strategies within the human health and biodiversity literature, presenting an opportunity to synthesize the two. We also explore common tradeoffs between these two goals, and recommend a series of design elements that can mitigate those tradeoffs.

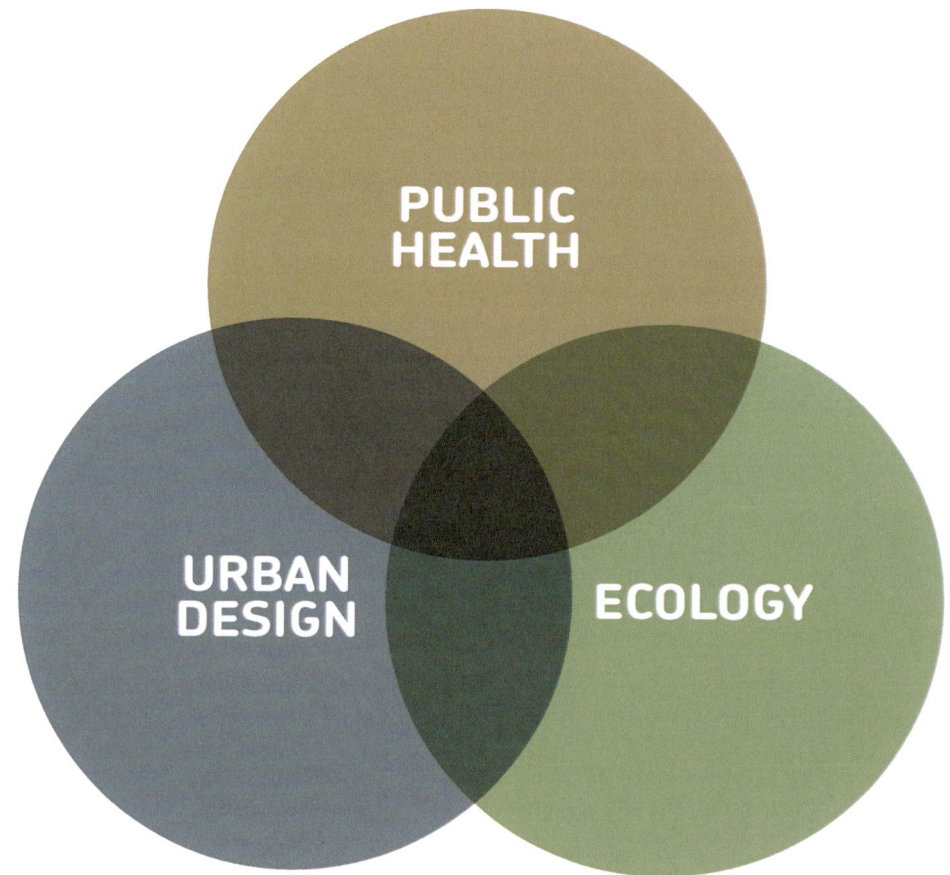

◀ Playground in Petaling Jaya, Malaysia (Gaddafi Rusli, Unsplash)

We developed strategies using the following principles:

1. **Ground recommendations in the scientific literature.**
Thorough review of the scientific literature was used to create recommendations that build on the current state of knowledge on urban greening, human health, and biodiversity. The scientific understanding of many of these concepts is still emerging, so we were careful to only speak to relationships that are established. We include extensive references throughout the text for readers to explore topics in greater detail.

2. **Seek solutions that support both human health and biodiversity, and be honest about tradeoffs.** In some cases the needs of native wildlife and people conflict. We worked to identify and address these conflicts and highlight options that can support both human health and native wildlife. We include a section within each element describing the potential tradeoffs between human health and biodiversity support, and include guidance to help reduce or resolve these tradeoffs wherever possible.

3. **Use design to explore tradeoffs and consequences**: In many cases, people benefit from access and exposure to greenspace, which can have negative impacts for wild plants and animals. We include a series of spatial design strategies in Chapter 3 to illustrate ways to resolve tradeoffs between human health and biodiversity support.

4. **Connect to existing efforts**. These strategies cover a wide array of complex topics, scales, site types, and disciplines, many of which have been described by experts in detail elsewhere. Wherever possible, we sought to elevate useful concepts through brief summaries and a list of key resources.

Bee on flowers (Jenna Lee, Unsplash) ▶

WHO IS SFEI?

The San Francisco Estuary Institute (SFEI) is a non-profit organization based in Richmond, California. For the past 30 years, SFEI has operated at the interface between science and policy, providing support for science-based environmental decision making by government agencies, private corporations, non-profit organizations, and communities. Through three major programs – Clean Water, Environmental Informatics, and Resilient Landscapes – SFEI aims to measurably improve the health and resiliency of ecosystems and communities in the San Francisco Bay Area and beyond.

For more than a decade, SFEI has been working to integrate the science of urban ecology into frameworks that allow planners and designers to support biodiversity in cities. In the course of this work, we noticed that many of our ideas for supporting urban biodiversity have parallel concepts in public health studies of human health benefits from urban greenspaces.

WHO IS RWJF?

This work is supported through a grant from the Robert Wood Johnson Foundation (RWJF). RWJF is the nation's largest philanthropy devoted to improving the health and well-being of everyone in America. In partnership with others, RWJF works to develop a Culture of Health rooted in equity. Please visit their website rwjf.org for more information.

DOCUMENT GUIDE

This document is organized into three chapters, each covering a different scale of intervention: planning, site, and detailed design (Figure 1). Planning strategies will be most relevant for those working on urban greening programs at the city or neighborhood level. The site and detailed design chapters are most relevant to decisions made by landscape architects and urban designers on individual greening projects. The site chapter strategies focus on layout and programming decisions that are specific to different types of urban greenspace, while design details are decisions that can be included in and are relevant to almost any urban greenspace. The three sections are tightly interrelated and include cross references to help the reader navigate between scales and topics. Throughout the document, connections to both larger and smaller scales are noted, with page references to allow the reader to quickly flip to a related concept.

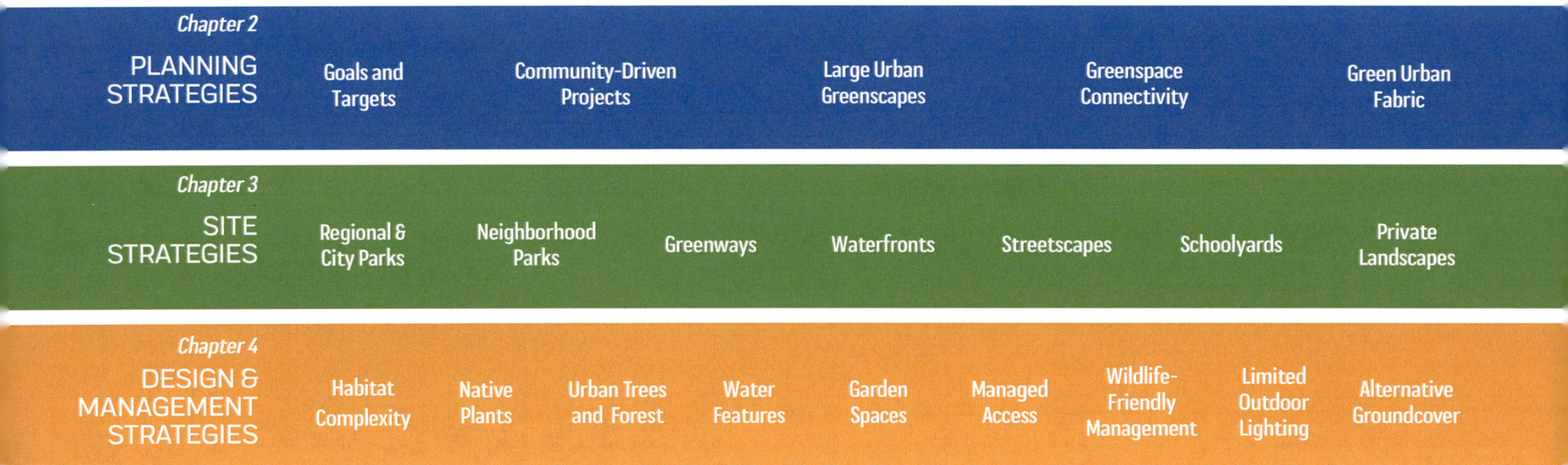

Chapter 2 **PLANNING STRATEGIES**	Goals and Targets	Community-Driven Projects	Large Urban Greenscapes	Greenspace Connectivity	Green Urban Fabric		

Chapter 3 **SITE STRATEGIES**	Regional & City Parks	Neighborhood Parks	Greenways	Waterfronts	Streetscapes	Schoolyards	Private Landscapes

Chapter 4 **DESIGN & MANAGEMENT STRATEGIES**	Habitat Complexity	Native Plants	Urban Trees and Forest	Water Features	Garden Spaces	Managed Access	Wildlife-Friendly Management	Limited Outdoor Lighting	Alternative Groundcover

Figure 1. Each of the planning, site, and design and management strategies operate at a different scale, and elements from each scale will likely apply to most sites.

Each strategy summarizes known benefits for human health and native wildlife, potential tradeoffs, and implementation guidance to resolve tradeoffs and maximize benefits. Site strategies feature sample designs illustrating key features for human health and biodiversity support. Finally, where relevant, the sections include key resources for further reading.

Given the complex and interwoven dynamics of ecosystems and urban spaces, reading through the entire document to learn about strategies from all three scales provides an understanding of biodiversity and human health support within cities. However, readers can also quickly look up strategies that are applicable to their program or project by identifying the relevant scale and strategy in the table of contents and skipping directly to the pages of interest.

WHAT IS BIODIVERSITY?

Biodiversity is a broad concept that refers to the variety of life on earth, including animal (vertebrates and invertebrates), plant, fungi, and microbial diversity. Design guidance within this document largely includes interventions to increase native plant diversity, which in turn supports increased diversity of animals, fungi, and microorganisms. *Wildlife* refers to the variety of native species in an environment including plants, animals, fungi, and microorganisms. *Ecology* refers to the study of these living species.

▲

(*Top*) Parliament buildings in Canberra, Australia (Mohit Kumar, Unsplash) (*Middle*) Swan Lake Open Space in London, UK (Landsil, Unsplash) (*Bottom*) California poppies in Southern California (Sergey Shmidt, Unsplash)

KEY CONCEPTS

While studying how landscapes can best support native wildlife, we found that designing cities for biodiversity in turn promotes human health. This document seeks to promote design solutions that can improve habitat parameters for native species, while maximizing health benefits of greenspaces for human communities.

Urban greening refers to the creation of *greenspace* within cities. Greenspace is broadly defined to include most areas that are not hardscape, including bare ground, herbaceous and shrub cover, trees, water features, and green roofs and walls. We consider greenspace within private landscapes, parks, and along streets. In some climates, greenspace may not be green for much or all of the year.

Why Urban Biodiversity?

In our increasingly urbanized world, the role cities play in supporting biodiversity has become vital. *Urban biodiversity* is the variety of living organisms present in a city. Within this document, biodiversity guidance focuses on ways that cities can support thriving communities of native wildlife, focusing mainly on vertebrates and insects. In addition to these groups, plant, fungal, and microbial diversity, as well as genetic diversity within species, should also be considered. This document targets support for *native wildlife*, which have coevolved in a particular geography, such that native species have specialized relationships with each other that create the diverse and dynamic natural heritage of a given location.

Overall, cities play a key role in global biodiversity conservation by providing critical habitat and acting as hotspots for threatened species.[1] In many cases, wildlife have no choice but to spend part of their time in cities, and quality urban greenspaces are key to providing them habitat.[2] Some *urban tolerant* species thrive in the unique conditions that cities offer, whether that be release from predators, additional food resources, or human tending. Other *urban intolerant* or *urban sensitive* species may move through cities to access more preferred areas, or may seek out habitat within large urban greenspaces buffered from human impacts.

This report documents the aspects of greenspaces that support the ability for wildlife to thrive[3,4] such as:

Habitat extent: Each species has particular needs for suitable habitat, including soil moisture, nutrients, and sun exposure for plants; availability of host plants for insects; nutrients for fungi; and appropriate breeding and foraging areas for birds. Wildlife species also vary in the size of habitat patches they require to thrive. Here we assume that greenspace area is a reasonable proxy for habitat extent.

Habitat quality: Habitat quality is a critical parameter defining how well an area can support a particular wildlife species. Presence of native plants and vegetation structure that matches historical vegetation communities can help boost habitat quality, as can wildlife-friendly management practices. Human disturbance can particularly limit the quality of *sensitive habitats*. For example, in wetlands, trampling can compact the soil and urban runoff can degrade water quality, resulting in poor habitat quality.

Habitat connectivity: Connectivity represents the ease with which mobile wildlife can move across the landscape, and can be measured as the distance between patches of greenspace and the absence of barriers blocking movement.

The guidance in this document is generalized to be true for most wildlife in most places. In a particular city, considering the needs of locally important species can help further refine these recommendations and provide more specificity for local projects. Selecting *focal species* with particular or representative habitat needs can help guide design decisions and support the development of goals and adaptive management strategies that support biodiversity overall.

❶ Flowers in Los Angeles, CA (Mark Hyde, Unsplash) ❷ Coyote (*Canus latrans*) in San Francisco, CA (Christopher Michel, Unsplash) ❸ Maggie Daley Park in Chicago, IL (Amie Bell, Unsplash)

(*Following pages*) ❹ Point Molate in Richmond, CA (SFEI) ❺ Keller Beach in Richmond, CA (SFEI) ❻ Coyote Hills Regional Park in Fremont, CA (SFEI) ❼ Brooklyn Bridge Park in NY, NY (SFEI) ❽ Designated wildlife habitat in Philadelphia, PA (SFEI) ❾ Brooklyn Bridge Park in NY, NY (SFEI)

Why Human Health?

Increasingly, research has shown links between greenspace and human health. Researchers have found three key pathways linking human health and well-being to greenspace access: reducing harm from environmental stressors or pollution; restoring capacities including cognitive function and impulse control; and building capacities including active living and social connections (Figure 2).[5] Research has established links through these pathways to improved physical and mental health, lowered disease, and reduced mortality for people with access to high quality greenspace.[5]

Many of the features of urban greenspace that support human health also foster biodiversity. For example, reduced harm and stress reduction for people is related to habitat complexity, which also means higher quality habitat for wildlife. Another example is that physical activity benefits for people relate to the size of connected greenspace, which relates to habitat extent for wildlife.

Furthermore, biodiversity *itself* supports human health and well-being. Studies are still sparse on this topic, but researchers have found that places with high species biodiversity support physical health,[6-8] improve positive mood and well-being,[9,10] and lower stress.[10-12]

Key Tensions and Tradeoffs

The tension between the needs of wildlife and the needs of people are real, and we highlight them throughout the document where they apply to a particular type of greenspace. Many of the trade-offs fall into these main categories:

Recreational access and disturbance: In places that both allow recreational use and attract wildlife, humans can disturb wildlife intentionally or unintentionally, potentially causing animals to expend energy being on alert or fleeing. When recreationists go off-trail, the level of disturbance to wildlife is heightened, and vegetation in sensitive areas gets trampled, degrading habitat quality.

Safety concerns: Residents may have safety concerns about wildlife. Whether a concern is actual or perceived, wildlife management and/or public education may be needed. Potential areas of concern include hygiene (e.g., animal droppings), disease (e.g., rabies and other zoonotic diseases), and human-wildlife interactions (e.g., protective behavior around nesting sites or young).

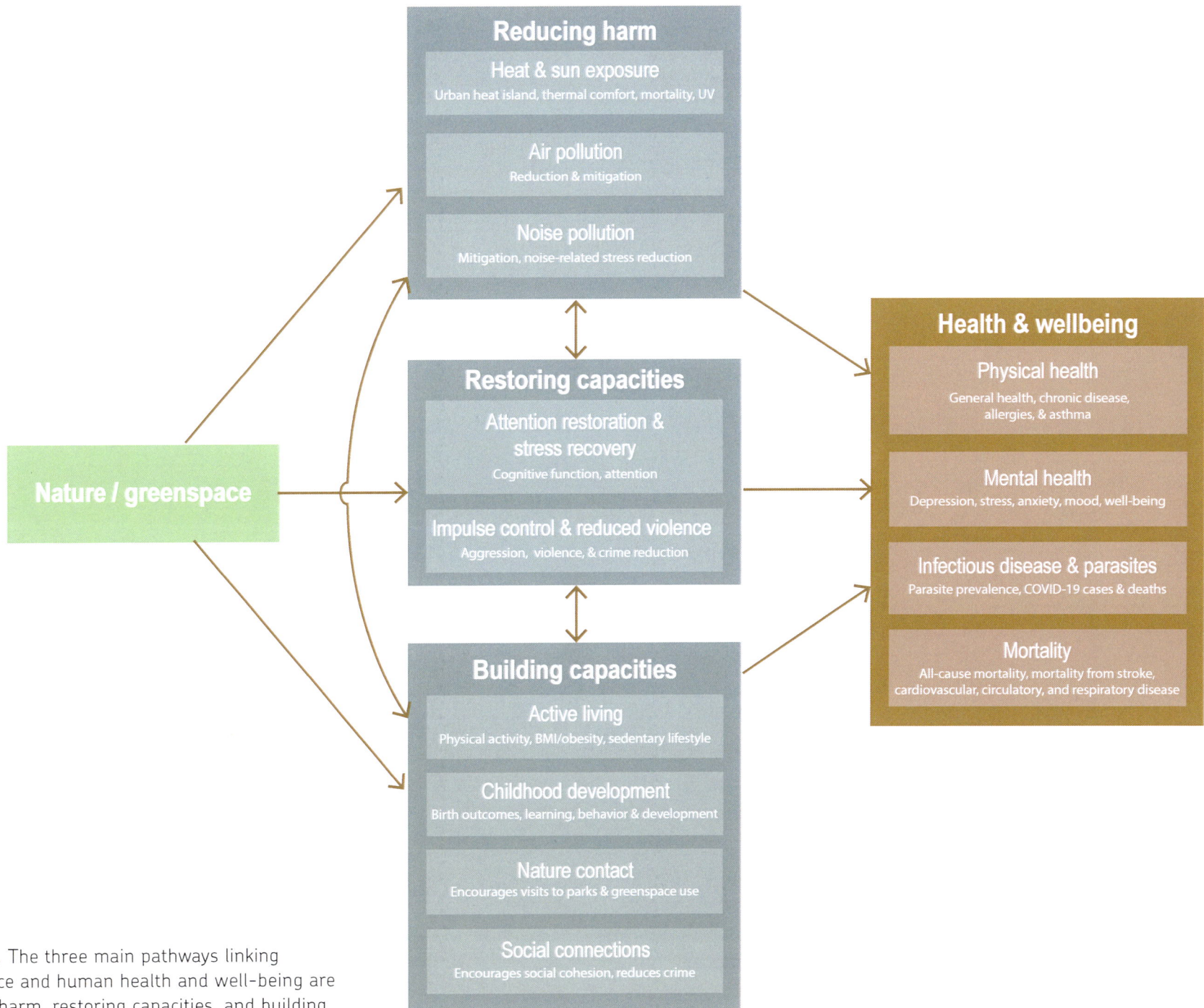

Reducing harm

Heat & sun exposure
Urban heat island, thermal comfort, mortality, UV

Air pollution
Reduction & mitigation

Noise pollution
Mitigation, noise-related stress reduction

Restoring capacities

Attention restoration & stress recovery
Cognitive function, attention

Impulse control & reduced violence
Aggression, violence, & crime reduction

Building capacities

Active living
Physical activity, BMI/obesity, sedentary lifestyle

Childhood development
Birth outcomes, learning, behavior & development

Nature contact
Encourages visits to parks & greenspace use

Social connections
Encourages social cohesion, reduces crime

Nature / greenspace

Health & wellbeing

Physical health
General health, chronic disease, allergies, & asthma

Mental health
Depression, stress, anxiety, mood, well-being

Infectious disease & parasites
Parasite prevalence, COVID-19 cases & deaths

Mortality
All-cause mortality, mortality from stroke, cardiovascular, circulatory, and respiratory disease

Figure 2. The three main pathways linking greenspace and human health and well-being are reducing harm, restoring capacities, and building capacities. Adapted from Marselle et al. (2021).[5]

ADDITIONAL RESOURCES

▶ Making Nature's City. https://www.makingnaturescity.org/

▶ RWJF. https://www.rwjf.org/

▶ Guide for Cities on Health-Oriented Planning and Use of Urban Green Spaces. 2022. UrbAct.

▶ Healing Gardens: Therapeutic Benefits and Design Recommendations. 1999. Marcus and Barnes.

Ecological traps: Urban greenspaces that are perceived as high quality habitat by wildlife, but do not support survival and/or reproduction, are known as *ecological traps*.[13] For example, planting a grove of fruit trees in a central business district may attract birds, who end up colliding with large reflective storefront windows.

Incompatible land use: Some human uses are best served by highly simplified greenspace, such as turf fields, which provide very little biodiversity value. Traditional western urban landscape forms tend to mimic savanna or forest communities with minimal shrubs or other understory. Even where people and wildlife may both benefit from the same cover type (e.g., forest), details of the design may differ. For example, dense vegetation that provides thick cover may be preferable for wildlife, while people may prefer open vegetation that allows clear views for safety.

Additional tensions and tradeoffs between these two potential objectives for urban greening are identified for each of the strategies presented in the subsequent chapters alongside design strategies to mitigate the tradeoffs.

CONSIDERATIONS AND LIMITATIONS

This document is limited by the current state of scientific knowledge. In particular, many of the effects of greenspace on human health are difficult to study because they are often confounded with other drivers, including poverty, discrimination, and education. In addition, we found that many studies evaluating public health outcomes did not provide specific information about the environmental or ecological systems in the study that could be used to guide design. While there is a strong body of evidence that overall greenness supports health, less is established on the benefits of specific types of site and design detail strategies. In comparison, the biodiversity literature provided much greater detail on site specific and design detail strategies and management implications. However, many studies on biodiversity outcomes were sparse on details of the built environment and social context. In some cases, this mismatch of information limited our ability to build connections between the two fields and suggest guidance. Further research is needed on the health benefits of site and design detail strategies such as street trees, structural complexity, and native plants. Studies that examine post-construction health benefits of site-scale design interventions can help to fill these gaps.

We acknowledge this guidance must be implemented according to the local social, economic, political, and ecological context. In particular, green gentrification is a nuanced, complex, and often site-specific topic crucial to consider in any discussion of urban greening. While we touch on the topic in this document, we encourage readers to further explore the topic through the expert resources we provide. See **Mitigating Green Gentrification (pg. 28)**.

This document synthesizes research on urban biodiversity and human health in cities across the globe, so many of the strategies can be applied in cities worldwide. However, our work is strongly informed by our experience as scientists and designers in California's San Francisco Bay Area. We acknowledge that this may result in some strategies and case studies being more relevant to the challenges of our region and those that share its economic, cultural, political, and ecological characteristics. We hope that this work can be expanded upon by ourselves and others in the future to increase its relevance to other parts of the world.

We hope to keep expanding our understanding of what strategies can support both biodiversity and human health in cities and are working with partners to help test and refine strategies in the San Francisco Bay Area. If you have additions, improvements, or even disagreements about these strategies, we would love to hear from you. Please reach out to: *EcologyforHealth@sfei.org*.

Ruby throated hummingbird (*Archilochus colubris*) in Canada (Jeremy Hynes, Unsplash) ▶

REFERENCES

1. Ives, C. D. et al. Cities are hotspots for threatened species: The importance of cities for threatened species. Glob. Ecol. Biogeogr. **25**, 117–126 (2016).

2. MacGregor-Fors, I. et al. City "green" contributions: the role of urban greenspaces as reservoirs for biodiversity. Forests **7**, 146 (2016).

3. Beninde, J., Veith, M. & Hochkirch, A. Biodiversity in cities needs space: a meta-analysis of factors determining intra-urban biodiversity variation. Ecol. Lett. **18**, 581–592 (2015).

4. Spotswood, E. et al. Making Nature's City: A Science-Based Framework for Building Urban Biodiversity. San Franc. Estuary Inst. Publ. (2019).

5. Marselle, M. R. et al. Pathways linking biodiversity to human health: A conceptual framework. Environ. Int. **150**, 106420 (2021).

6. Prescott, S. L. et al. The skin microbiome: impact of modern environments on skin ecology, barrier integrity, and systemic immune programming. World Allergy Organ. J. **10**, 1–16 (2017).

7. Prescott, S. L., Logan, A. C., Millstein, R. A. & Katszman, M. A. Biodiversity, the human microbiome and mental health: moving toward a new clinical ecology for the 21st Century. Int J Biodivers. **2016**, 1–18 (2016).

8. Hanski, I. et al. Environmental biodiversity, human microbiota, and allergy are interrelated. Proc. Natl. Acad. Sci. **109**, 8334–8339 (2012).

9. Cameron, R. W. et al. Where the wild things are! Do urban green spaces with greater avian biodiversity promote more positive emotions in humans? Urban Ecosyst. **23**, 301–317 (2020).

10. Fuller, R. A., Irvine, K. N., Devine-Wright, P., Warren, P. H. & Gaston, K. J. Psychological benefits of greenspace increase with biodiversity. Biol. Lett. **3**, 390–394 (2007).

11. Schebella, M. F., Weber, D., Schultz, L. & Weinstein, P. The wellbeing benefits associated with perceived and measured biodiversity in Australian urban green spaces. Sustainability **11**, 802 (2019).

12. van den Bosch, M. & Ode Sang, Å. Urban natural environments as nature-based solutions for improved public health - A systematic review of reviews. Environ. Res. **158**, 373–384 (2017).

13. Robertson, B. A. & Hutto, R. L. A framework for understanding ecological traps and an evaluation of existing evidence. Ecology **87**, 1075–1085 (2006).

◀ Two-tailed swallowtail (*Papilio multicaudata*) in Mexico City, Mexico (Don Fabia, Unsplash)

02
PLANNING
STRATEGIES

PLANNING STRATEGIES

This chapter focuses on key planning strategies that can be used to create cities that support both biodiversity and human health. Planning strategies set the stage for site scale activities by aligning goals, identifying opportunities, and setting priorities. This alignment supports systems that work together, for both people and wildlife. These strategies can help coordinate efforts from the regional to neighborhood scale and inform the implementation of the site and design detail-based strategies discussed in the following chapters. The first two strategies focus on process-based approaches: setting appropriate **Urban Greening Goals and Targets (pg. 18)** and incorporating **Community-Driven Projects (pg. 20)**. The latter three strategies relate to the spatial distribution of greening across the urban landscape: **Large Urban Greenspaces (pg. 22)**, **Greenspace Connectivity (pg. 24)**, and a **Green Urban Fabric (pg. 26)**.

Typical contexts and documents in which these strategies may be relevant include:

▶ City-wide planning (e.g., General Plans, Parks or Open-Space Master Plans, Urban Forest Master Plans, Biodiversity Strategies, etc.)

▶ Community engagement efforts for plan and project development

▶ City programs (e.g., residential incentive programs)

▶ Neighborhood planning (e.g., Specific Plans)

◀ Green space and playground (Nerea Martí Sesarino, Unsplash)

URBAN GREENING GOALS AND TARGETS

Goals and targets are an explicit description of needs for urban greening that can help direct future actions and priorities. Here, urban greening goals include formalized citywide or regional strategies and planning documents as well as future visions. Urban greening targets are specific quantitative measures associated with those goals.

Biodiversity and Human Health Benefits: Goals and targets help cities achieve the biodiversity and health benefits of greening. Goals and targets enable cities to prioritize projects, track progress, and adjust as necessary. They can improve coordination and access to funding opportunities across projects and departments. They can also help managers evaluate and balance any necessary tradeoffs between human health benefits and supporting native wildlife.

IMPLEMENTATION GUIDANCE

▶ **Be specific.** To improve likelihood of success, include clear goals, define pathways to achieve those goals, and identify resources to ensure the ongoing success of the plan. Greater specificity here enables clear communication across government agencies, partner organizations, and the general public. It also supports resource allocation and adaptive management. Specific quantitative targets also enable progress to be measured, supporting adaptive management. Identify whether each goal is intended to be prescriptive (related to something that can be controlled, such as number of trees planted) or performance-based (related to the desired final outcome, such as an increase in bird diversity or number of people exercising).

▶ **Establish priorities.** Because urban greening goals can be expensive and compete against a range of other needs in urban policy and design, specifying priorities is important to achieve success. Potential focuses for urban greening include: protecting endemic species, preserving or enhancing regional and global migration corridors, fostering ecosystem services, providing equitable access and creating a sense of place, and protecting culturally and historically important species and sites.

▶ **Set biodiversity goals and targets.** Biodiversity goals and targets may focus on either actions to support biodiversity (e.g., habitat diversity), or the biodiversity response (e.g., species richness).

- Identify focal/respresentative species: Identify a set of representative species to help guide actions for restoration and protection. Species vary in the size of patch and types of cover they require as habitat (e.g., some beetle species require open ground for reproduction, while most mammals cross bare ground to access patches of green). Species also differ in the distance of non-habitat that they can successfully traverse. For more information on patch size and spacing see **Greenspace Connectivity (pg. 24)**. Working to address the needs of a few representative species can make discussion of targets and tradeoffs more concrete and can focus the effort on feasible interventions.

- Multi-species strategies: Different species have different requirements for surviving in the urban environment, and design should work to meet the needs of multiple species. Using focal species, consider how each habitat patch or corridor can support multiple modes of animal movement and plant dispersal (e.g., for animals: flight, walking, swimming, arboreal; for plant seed dispersal: air, water, hitch-hiking on animals).

- <u>Habitat function.</u> To ensure that a habitat is providing the desired biodiversity support and not functioning as an *ecological trap*, consider goals and targets that represent habitat function, including primary productivity, connectivity, or successful reproduction of target species.

▶ **Set human health goals and targets.** Although the link between any specific landscape change and a human health outcome is hard to measure, many human health outcomes improve with increased access to greenspace. Examples of prescriptive targets include percent of population within walking distance of a park or percent tree canopy cover by neighborhood. Outcome-based goals or targets could include an increase in usage or improved self-reported health status.[1] Tracking targets across the entire urban area can identify areas most in need of greenspace, helping to ensure equitable distribution of benefits. Involve the community in identifying specific health concerns and goals that are locally relevant. See **Community-Driven Projects (pg. 20)**.

▶ **Identify co-benefits.** Setting goals for multiple benefits within a strategy can lead to a higher return on investment and increased opportunities for funding and coalition building. Strategies that support biodiversity and health may also contribute to other goals including: stormwater management, mitigating urban heat island effect, community building, job creation, and climate change mitigation and adaptation.

▶ **Acknowledge tradeoffs.** In some situations, actions to support native wildlife and human health may not be compatible. Outside of very large parks, apex predators such as bears, coyotes, and mountain lions may not be compatible with human use. In small spaces, people may most easily share with birds and insects. Clearly establishing biodiversity and health goals presents an opportunity to acknowledge and specify tradeoffs.

▶ **Use adaptive management.** A clear set of goals and targets supports adaptive management. Conduct regular monitoring to assess progress towards goals and make adjustments in strategies as needed. If monitoring identifies a lack of progress or uneven progress towards goals, consider changing strategies, re-engaging with communities, improving the specificity of goals and targets, or changing the scale of intervention.

❶ Park playground (Honey Yanibel, Unsplash) ❷ San Francisco Bay (SFEI) ❸ Rose Kennedy Greenway in Boston, MA (Kindra Clineff, CC BY 4.0) ❹ Park in Tokyo, Japan (Jakub Dziubak, Unsplash)

COMMUNITY-DRIVEN PROJECTS

Urban planning and design typically involve community engagement, which is the voluntary participation of local residents in certain parts of the planning and design process. Community-driven projects emphasize equitable engagement that begins early in the process and provides benefits to the local residents at the project's completion. Community-driven design is a key strategy for combating green gentrification (see page 28).

Biodiversity Benefits: Community involvement in ongoing stewardship has been shown to improve the success rate of greening projects (e.g., a higher survival rate of planted trees).[3,4]

Human Health Benefits: Greening efforts that meet the needs of the surrounding community may be more likely to be used by residents and improve health outcomes for underserved residents.

KEY TENSIONS

▶ Productive engagement between community members and biodiversity advocates is often time consuming. If adequate time is not invested building a shared understanding of needs and constraints, opportunities may be missed for projects that meet multiple needs.

▶ Real tradeoffs often exist between community and biodiversity goals, and balancing different needs can be challenging. In some cases, other land uses, such as affordable housing, may be prioritized over additional greenspaces. At the design level, preferences for urban greenspace elements may also limit the ability of a site to support wildlife.

▶ Although urban greenspace contributes to human health, it has a relatively modest impact compared to other *social determinants of health* such as access to education, health care, clean water, and healthy food. For cities with limited budgets and pressing public health concerns, urban greening may not be the most cost-effective way to support resident health, even considering the additional benefits to native wildlife.

IMPLEMENTATION GUIDANCE

▶ **Plan community engagement intentionally.**

- Start early. Begin engaging community members as early as possible in the process, including developing partnerships and fundraising opportunities together with community members before identifying the site and the scope of the project. This engagement allows for collaborative ideation and budgeting, creation of a shared and more complete understanding of opportunities and constraints, and inclusion of community leadership in the team.

- Invite broad, inclusive participation in the process, aiming to achieve equitable representation by impacted communities. Consider ways to make content and participation accessible to a diverse range of audiences who may have a range of barriers, e.g., cultural, socioeconomic, ability/disability, or language differences. Identify logistical barriers to participation, and ways to mitigate those barriers, for example, scheduling meetings at times when community members can attend, providing food and/or childcare during meetings, or holding in-person events at an ADA-compliant venue that is accessible by public transportation. Acknowledge the value of participation by compensating community members for their time and expertise.

- Build communication. If the community has not been engaged prior to the conceptual design process, hold a project kickoff meeting to introduce the community to the design team and purpose of the project. Ask the community members to share who they are, where they live, what their needs are, and how they use the space if a site has been identified. Be willing to change the project site and scope based on community feedback.

▶ **Engage actively.** Facilitators should pursue a genuine and robust community engagement process. Engagements that are not community-driven or do not center social equity should not be labeled as such. These types of unsubstantiated or misleading claims, whether made intentionally or unintentionally, have the potential to cause harm and break down trust.

▶ **Feature community-driven design.** Design urban greening projects and their associated features, amenities, and programming to celebrate local culture and address community needs and interests. Community meetings

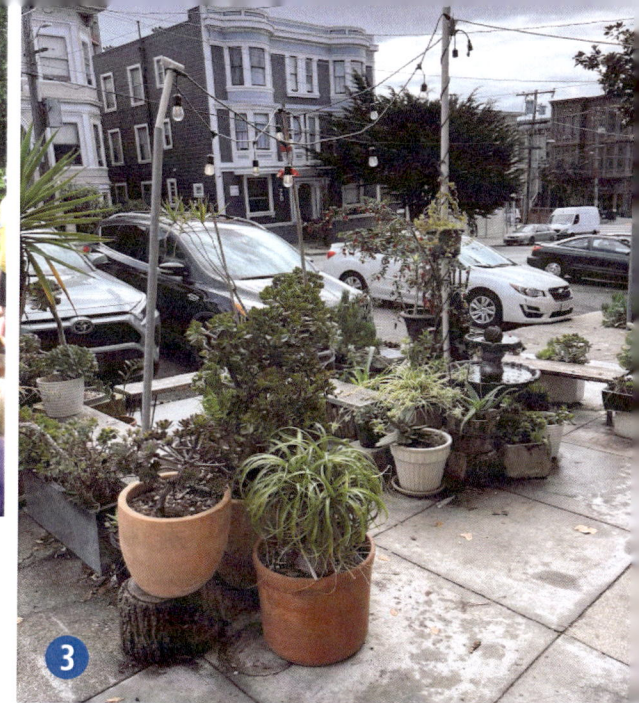

during the conceptual and schematic design phases can be used to get feedback to ensure the space matches what the community wants and offers the benefits they are interested in (exercise, quiet space for reflection, nature, etc).

▶ **Involve the community in implementation.**

- Source construction labor and material locally when possible so that project funds stay in the community.[5] Depending on the process, community design-build projects can help foster a sense of place. For example, Pogo Park in Richmond is a revitalization of an existing public park that was community-led. It was designed and partially constructed by community members.[6]

- Encourage volunteer participation to engage the community and create a sense of ownership of the space. Examples include restoration planting events in habitat areas and community art projects.

▶ **Seek continued engagement.**

- Establish programs to monitor progress towards the project goals and provide ongoing stewardship of the completed project to ensure it continues to provide intended benefits. Monitoring and stewardship programs are an opportunity to provide job-training and actual employment to residents and engage youth.[5] Stewardship programs can support workforce development for youth and adults by providing hands-on experience, job-training, and a pipeline to long-term, well-paying jobs.

- Reduce barriers such as park fees, rules, and limited operating hours to increase accessibility and equity for park users.[7] Include programming such as concerts, tai chi, potlucks, and sports events to support park use and recreation.[7] Maintaining recreational facilities and greenspaces supports physical activity[8] and perceptions of safety.[9]

ADDITIONAL RESOURCES

▶ International Association of Public Participation. https://www.iap2.org/page/resources.

1 Parking lot plantings in Mountain View, CA (SFEI) **2** Tree planting in Virginia (Albert Herring) **3** Steet plantings in San Francisco, CA (SFEI) **4** Community garden in Leiden, Netherlands (John Lord, CC BY 4.0)

LARGE URBAN GREENSPACES

Large greenspaces, defined here as being larger than 5 acres (~2 hectares), can provide some of the most substantial support for native wildlife and human recreation within cities and should be a priority component of urban greening efforts. These greenspaces are often public amenities such as nature reserves and large parks that serve entire regions or cities. For site-scale design strategies for these spaces, see Regional and City Parks (pg. 34).

Biodiversity Benefits: Habitat patches larger than 5 acres (~2 hectares) are essential for many species, particularly wildlife that are sensitive to urban impacts.[10] Large greenspaces tend to have more types of habitats, including those that provide for foraging, nesting, and protection from predators. They are also large enough to include buffer areas between core habitat and roadways, impervious surfaces, and other urban hazards.

Human Health Benefits: Large parks have many human health benefits. Parks greater than 5 acres (~2 hectares) have been found to support more physical activity, including walking.[11,12] In New York City, researchers found that proximity to large parks greater than 6 acres was significantly associated with lower BMI (body mass index) while proximity to small parks was not associated with lower BMI.[13] Larger parks also have stronger cooling effects.[14]

KEY TENSIONS

▶ The availability of large habitat patches strongly influences the number and variety of species that can thrive in a city. By contrast, some human health measures respond most to proximity to parks rather than size, and to achieve equitable access to greening, more distributed greenspaces may be preferable to a few large parks.

▶ There is some evidence that large parks are more likely to lead to green gentrification than smaller parks.[15,16] However, other studies have found unclear or absent relationships between park size and gentrification.[17,18]

IMPLEMENTATION GUIDANCE

▶ **Engage communities.** Different communities have distinct needs for greenspaces, and the combination of large and distributed greenspaces that best support community and biodiversity needs will vary. Community engagement is also a key strategy for combating gentrification. See **Community-Driven Projects (pg. 20)**.

▶ **Protect existing habitat in cities**. Intact natural habitat areas typically provide much better support for native wildlife than restored areas,[19,20] and creating new habitat can be an expensive and slow process. Preserve remnant patches of habitats in cities when possible.[21]

▶ **Create parks on brownfield sites.** Opportunities for creating new parks in developed areas are often limited by the availability of appropriate sites. Disturbed *brownfield sites*, or areas previously developed which are no longer in use, are often the only land available for new large parks.[22] While careful remediation of the site is needed, many popular parks such as Tempelhof Field in Berlin, Gas Works Park in Seattle, and Freshkills Park in New York have been built on reclaimed sites.

▶ **Prioritize regional biodiversity hubs.** Create and manage greenspaces larger than 125 acres (~50 hectares) as regional hubs for both biodiversity and human use. Urban intolerant species and other more sensitive wildlife are more likely to be attracted to very large parks that offer varied habitats with minimal human disturbances.[23] A recent international review of 75 cities found that greenspaces of 125 acres (50 hectares) or more were necessary to conserve threatened or urban intolerant wildlife.[10]

▶ **Choose shapes that maximize interior habitat.** Design parks that have a shape that is compact, such as a square or circle, rather than long and narrow to reduce the edge area of the park.[24] Parks with lower perimeter-to-area ratios have cooler interiors[25,26] and are less impacted by edge effects such as noise and light that can be detrimental to wildlife.[10,27]

▶ **Site large parks near people.** Where possible given site availability, locate large parks to maximize the number of residences within half a mile from the park to ensure that people benefit from parks.[13] Prioritize investment in historically underserved communities with limited access to quality greenspaces while also considering the potential for green gentrification. See **Mitigating Green Gentrification (pg. 28).** Where large greenspaces are not feasible, consider the creation of distributed features that can provide benefits for people and enhance connectivity. See **Greenspace Connectivity (pg. 24).**

❶ Gas Works Park in Seattle, WA (CC BY 4.0)
❷ Central Park in NY, NY (Hector Argüello Canals, Unsplash) ❸ Bay Trail in Hayward, CA (SFEI) ❹ Oro Loma Marsh in Hayward, CA (SFEI)

GREENSPACE CONNECTIVITY

Publicly accessible greenspaces within a city can be strategically placed and linked to create a greenspace network that distributes access for residents across the city and helps wildlife move safely between patches. Key components of a greenspace network can include corridors of continuous greenspace and distributed small parks that serve neighborhoods and function as stepping stone habitat patches for wildlife. For more information on benefits and strategies for encouraging urban greening in areas outside the formal park network, such as streets and private residences, see Green Urban Fabric (pg 26).

Biodiversity Benefits: Greenspace connectivity supports wildlife movement across fragmented urban habitats. Corridors (such as greenways) and stepping stones (such as small greenspaces) between large habitat patches enable wildlife to move across the landscape and help reduce species loss in cities,[28] particularly for species that are intolerant of high levels of urbanization.[29]

Human Health Benefits: A distributed network of greenspaces near residences improves park access and is thought to increase park use and physical activity.[30,31] Having parks and other greenspaces near residences is associated with improved mental health, life satisfaction, and social functioning, as well as reduced BMI, all-cause mortality, and likelihood of diabetes, sleep disorders, and joint disease.[32,33] For children, having a park nearby improves physical health, connection with nature, behavior and emotional development, and lowers hyperactivity/inattention.[34,35]

KEY TENSIONS

▶ Limited budgets for creating and maintaining parks require choices in priorities. Wildlife benefits most from greenspaces that are concentrated near other large green patches which allows for movement between key resources.[28,36] By contrast, human health may be best served by focusing on disadvantaged neighborhoods that have been historically deprived of natural space and by creating distributed greenspaces that maximize access.[37,38]

IMPLEMENTATION GUIDANCE

▶ **Determine landscape-scale greenspace needs for wildlife and people**. For wildlife, identify parks and patches that are far away from other patches or poorly connected to other greenspaces. To identify barriers and gaps in the network, the preferences of focal species can be used to develop distance thresholds between habitat patches and to perform connectivity analyses.[39,40] For people, use community engagement and spatial analysis to identify areas with limited human access to greenspaces, especially around underserved communities. Engagement with communities is necessary to understand local greenspace needs and barriers to use. Common spatial metrics used for human connectivity analyses include minimum distance to parks, walking times, and quality and coverage of active transit infrastructure such as cycling routes, multi-use paths, trails, and green streets.

▶ **Prioritize locations that can boost both human access and connectivity for wildlife.** Site parks and greenways in locations that could both improve equity in human access and enhance connectivity for wildlife. When space for connectivity is limited for a particular site, consider using site-scale strategies to reduce conflict between these uses. See **Neighborhood Parks (pg. 36)**, **Greenways (pg. 38)**, and **Waterfronts (pg. 40)**.

▶ **Prioritize locations near transit and neighborhoods historically deprived of greenspace.** Positive predictors of park visitation include length of bike routes, number of subway stops and proximity to nearest bike route, bus stop and subway stop.[41]

▶ **Identify barriers to wildlife movement such as major roadways, highways, and densely developed areas.** Where these barriers block important habitat access, improve connectivity through designing corridors or stepping stones to support safe wildlife movement. Reduce gaps between greenspaces by creating connections, such as greenways, trails, landscaped cul-de-sacs, and green alleys to encourage wildlife movement.

1 **Waterfronts**
are strips of land along a water body that can provide recreational opportunities while supporting unique species assemblages due to their proximity to water

2 **Neighborhood parks**
are typically less than 5 acres (~2 hectares) and are recommended to be distributed within 0.3 miles (500m) of each other to increase human access to greenspaces

3 **Greenways**
are relatively narrow linear greenspaces that usually include multi-use pathways for active transportation (walking, biking, etc.) in addition to vegetated areas

4 **Focal species**
are key wildlife species that can be studied to develop distance thresholds for identifying gaps and barriers in the greenspace network

5 **Stepping stones**
are small greenspaces that enable wildlife to move across the landscape and help reduce species loss in cities

6 **Regional and city parks**
are defined as being at least 5 acres (~2 hectares) in size. They typically serve either the entirety or significant portions of a city or region

▶ **Selectively design for people or biodiversity.** Where biodiversity and human needs do not align, design for the target benefit. For example, in areas that lack greenspace for human use, create a distributed network of small neighborhood parks within 0.3 miles (~500 meters) to increase human health benefits associated with good access to greenspaces.[13,33] In areas that have lower human need for improved access but high biodiversity need for connectivity, prioritize creating continuous corridors or larger stepping stones that lead to larger greenspaces[10,28] and choose plantings that will support wildlife.

▶ **Daylight streams.** Daylight buried streams to restore habitat and accessibility to the waterway.[42] Streams support unique wildlife assemblages, and can function as important connectivity features on the landscape.[43,44] Daylighting creeks also creates new opportunities for the public to interact with water. See Waterfronts (pg. 40).

GREEN URBAN FABRIC

In ecology, the matrix is the area in which habitat patches are embedded. In an urban context, this is typically everything outside of formal park spaces and green corridors. Here, we use the term *green urban fabric* to represent the potential for the urban matrix to have vegetation and greenness woven throughout. Opportunities include vegetated street buffers, green vacant lots, green stormwater infrastructure, parking lot trees, office parks, campuses, gardens, and private yards.[47] For more information on planning networks of formal parks, see Greenspace Connectivity (pg 24).

Biodiversity Benefits: A robust green urban fabric can provide added connectivity for animals moving between park patches and provide habitat and food resources for smaller urban-adapted animals such as pollinator insects, small mammals, and birds.[48,49] A greener urban fabric can increase the effective size of patches, so that they can support more species.[50]

Human Health Benefits: Non-park components of an urban landscape can significantly increase greenness, often measured using remote sensing to measure the percent of an area covered by vegetation. Living and working in areas with higher overall greenness have been associated with numerous positive health outcomes, including greater physical activity, lower obesity rates, and better mental health.[51,52] A green urban fabric can also mitigate urban heat island effect.[53]

KEY TENSIONS

▶ Small greenspaces within the urban matrix may not be able to support both human and biodiversity uses. Health benefits are typically associated with measures such as total greenness, which may not necessarily support biodiversity, especially if plantings are non-native and lack complex structure. For example, high-use pocket parks or street buffers in busy areas may provide value for people but the level of human disturbance may limit use by wildlife.

▶ Small greenspaces that are not well-connected to habitat patches have the potential to function as ecological traps, which lower overall survival or fitness for wildlife. Small greenspaces may expose wildlife to predation, roadways, and other stressors, with little potential to escape.

IMPLEMENTATION GUIDANCE

▶ **Integrate private spaces with public greenspace networks.** Privately owned, publicly accessible greenspaces play an increasing role in the greenspace networks of many cities. Including these spaces in local plans such as specific and area plans can help ensure that these private spaces are connected with public greenspaces. Educational and incentive programs can encourage landowners to create or improve private greenspaces.

▶ **Prioritize connections.** For wildlife, a green urban fabric is most helpful when it connects patches of habitat and makes it safer to move across the landscape. Non-park greenspaces can serve as stepping stones for dispersal or can augment the functional value of adjacent parkland.

▶ **Increase tree canopy cover.** Programs to both maintain existing street trees and encourage additional tree planting can help increase the urban tree canopy. Block-scale tree canopy cover of 40% has been shown to support urban wildlife and reduce impacts from the urban heat island effect,[54] but may not be achievable in some contexts. The Tree Equity Score tool can be used to set achievable targets based on biome and urban density in U.S. cities.[55] Parking lots and street margins are often relatively easy places to plant trees for public benefits. In arid regions the benefits of tree planting should be weighed against the increased water use. See **Urban Trees and Forests (pg. 58).**

▶ **Use native plants.** Connect habitat patches with native plantings to allow wildlife to move through and access different resources.[56] Widespread use of common non-native plants exacerbates fragmentation by homogenizing vegetation and reducing the potential for multiple distinct habitat types to occur across urban landscapes.[57,58] Also, native plants often better support native insects and wildlife, due to their coevolutionary relationships. See **Native Plants (pg. 56).**

▶ **Engage homeowners.** In cities with many single-family homes and suburban development, p vately owned yards present an opportunity to create additional habitat.[59] Educational campaigns and resources for homeowners can help create understanding of the potential benefits of these spaces and encourage participation. See **Private Landscapes (pg. 46)** and **Garden Spaces (pg. 62)**.

▶ **Convert vacant lots.** Vacant lots provide an excellent opportunity for cities to create small greenspaces with benefits for biodiversity and people.[60] One study in Philadelphia found vacant lot conversion resulted in a 76% increase in time spent outdoors and a 29% decline in local shootings.[9] Vacant lots contribute to urban avian diversity and can serve as valuable tools in urban conservation.[60]

❶ Green infrastructure in El Cerrito, CA (SFEI) ❷ Street garden in San Francisco, CA (SFEI) ❸ Davie Community Garden in Vancouver, BC, Canada (Geoff Peters, Unsplash) ❹ Trees in Washington D.C. (Ted Eytan, CC BY 4.0)

MITIGATING GREEN GENTRIFICATION

Gentrification is the process that is triggered by planning and investments in a historically disinvested neighborhood, which leads to new interest in those neighborhoods from outside investors. This interest drives up land values and the cost of living, which can result in economic hardship for and/or displacement of the original residents. Typically the neighborhood demographics shift, as the investments in the neighborhood attract new residents who tend to be white, have more college degrees, and have higher income. Residents of neighborhoods at risk of gentrification tend to have lower income, rent their homes, and have fewer college degrees. Green gentrification is the same process, triggered specifically by planning and investments for urban greening projects.

Recent research focused on identifying which spatial, physical, and institutional characteristics of greening efforts lead to green gentrification has found that some greening actions, such as large-scale infrastructure projects to create new large parks or greenways, may be more likely to trigger green gentrification than small scale improvements such as tree planting.[18,61,62] While some studies have found evidence that larger parks may trigger green gentrification more than smaller parks,[15,16] others have found unclear[17] or no links to park size.[18]

In order to successfully distribute the benefits of urban greening across neighborhoods and avoid green gentrification, cities must adopt a multi-disciplinary, multi-pronged approach that is tailored to the unique needs and perspectives of the community. A best practice is to start early, well before any planning for urban greening is considered and before outside investors take interest in the neighborhood. Experts suggest a variety of strategies, a small sample of which are listed below.

- **Tenant protections:** Agencies can enact policies that stabilize housing for renters by limiting the allowable annual rent increases and protecting tenants from unjust evictions.

- **Land use and housing:** Agencies can require and/or incentivize building of more affordable housing units.

- **Align funding with equitable outcomes:** Agencies that distribute funds for greening projects can establish equity as a criterion for funding proposals.

- **Enhance economic opportunities:** Provide job training to and prioritize hiring of local residents so that local economic opportunities grow as a result of greening investments.

For more information, refer to the following resources:

Gibbons, A., et al. Greening in place: Protecting communities from displacement. Audubon Center at Debs Park, Public Counsel, SEACA-LA, and Team Friday (2020). https://www.greeninginplace.com/s/GG-2020-ToolKit-FINAL.pdf

Mohnot, S., J. Bishop, and A. Sanchez. "Making Equity Real in Climate Adaptation and Community Resilience Policies and Programs: A Guidebook." The Greenlining Institute: Oakland, CA, USA (2019). https://greenlining.org/wp-content/uploads/2019/08/Making-Equity-Real-in-Climate-Adaption-and-Community-Resilience-Policies-and-Programs-A-Guidebook-1.pdf

Rigolon, A., and J. Christensen. "Greening without gentrification: Learning from parks-related anti-displacement strategies nationwide." UCLA: Los Angeles, CA, USA (2019). https://www.ioes.ucla.edu/wp-content/uploads/Greening-without-Gentrification-report-2019.pdf

Oscilowicz, E. et al. Policy and Planning Tools for Urban Green Justice. C40 Knowledge (2021) https://www.c40knowledgehub.org/s/article/Policy-and-Planning-Tools-for-Urban-Green-Justice?language=en_US

REFERENCES

1. Gehl Institute. *Inclusive Healthy Places*. (2018).

2. Mata, L. *et al.* Punching above their weight: the ecological and social benefits of pop-up parks. *Front. Ecol. Environ.* **17**, 341–347 (2019).

3. Boyce, S. It Takes a Stewardship Village: Effect of Volunteer Tree Stewardship on Urban Street Tree Mortality Rates. *Cities Environ.* **3**, 1–8 (2010).

4. Roman, L. A. *et al.* Stewardship matters: Case studies in establishment success of urban trees. *Urban For. Urban Green.* **14**, 1174–1182 (2015).

5. Gibbons, A. *et al. Greening in Place: Protecting Communities from Displacement*. https://static1.squarespace.com/static/5f5ab412f824d83e0eefa35e/t/5f739385c6cc3d63acd8d875/1601409949612/GG-2020-ToolKit-FINAL.pdf (2020).

6. Pogo Park. Pogo Park Who We Are. *Pogo Park* https://pogopark.org/who-we-are/history/ (2022).

7. Kuo, M. Nature-deficit disorder: Evidence, dosage, and treatment. *J. Policy Res. Tour. Leis. Events* **5**, 172–186 (2013).

8. de Vries, S. I., Bakker, I., van Mechelen, W. & Hopman-Rock, M. Determinants of activity-friendly neighborhoods for children: results from the SPACE study. *Am. J. Health Promot. AJHP* **21**, 312–316 (2007).

9. Branas, C. C. *et al.* Citywide cluster randomized trial to restore blighted vacant land and its effects on violence, crime, and fear. *Proc. Natl. Acad. Sci.* **115**, 2946–2951 (2018).

10. Beninde, J., Veith, M. & Hochkirch, A. Biodiversity in cities needs space: a meta-analysis of factors determining intra-urban biodiversity variation. *Ecol. Lett.* **18**, 581–592 (2015).

11. Koohsari, M. J. *et al.* Are public open space attributes associated with walking and depression? *Cities* **74**, 119–125 (2018).

12. Paquet, C. *et al.* Are accessibility and characteristics of public open spaces associated with a better cardiometabolic health? *Landsc. Urban Plan.* **118**, 70–78 (2013).

13. Rundle, A. *et al.* Associations between Body Mass Index and Park Proximity, Size, Cleanliness, and Recreational Facilities. *Am. J. Health Promot.* **27**, 262–269 (2013).

14. Cao, X., Onishi, A., Chen, J. & Imura, H. Quantifying the cool island intensity of urban parks using ASTER and IKONOS data. *Landsc. Urban Plan.* **96**, 224–231 (2010).

15. Chen, Y. *et al.* Can smaller parks limit green gentrification? Insights from Hangzhou, China. *Urban For. Urban Green.* **59**, 127009 (2021).

16. Kim, S. K. & Wu, L. Do the characteristics of new green space contribute to gentrification? *Urban Stud.* **59**, 360–380 (2022).

17. Wu, L. & Rowe, P. G. Green space progress or paradox: Identifying green space associated gentrification in Beijing. *Landsc. Urban Plan.* **219**, 104321 (2022).

18. Rigolon, A. & Németh, J. Green gentrification or 'just green enough': Do park location, size and function affect whether a place gentrifies or not? *Urban Stud.* **57**, 402–420 (2020).

19. Soga, M., Yamaura, Y., Koike, S. & Gaston, K. J. Woodland remnants as an urban wildlife refuge: a cross-taxonomic assessment. *Biodivers. Conserv.* **23**, 649–659 (2014).

20. Jones, H. P. *et al.* Restoration and repair of Earth's damaged ecosystems. *Proc. R. Soc. B Biol. Sci.* **285**, 20172577 (2018).

21. World Economic Forum & Alexander von Humboldt Institute. *BiodiverCities by 2030: Transforming Cities' Relationship with Nature*. https://www3.weforum.org/docs/WEF_BiodiverCities_by_2030_2022.pdf (2022).

22. Czerniak, J. *Large Parks*. (Princeton Architectural Press, 2007).

23. Collinge, S. K. Ecological consequences of habitat fragmentation: implications for landscape architecture and planning. *Landsc. Urban Plan.* **36**, 59–77 (1996).

24. Matthies, S. A., Rüter, S., Prasse, R. & Schaarschmidt, F. Factors driving the vascular plant species richness in urban green spaces: Using a multivariable approach. *Landsc. Urban Plan.* **134**, 177–187 (2015).

25. Ren, Z. *et al.* Estimation of the relationship between urban park characteristics and park cool island intensity by remote sensing data and field measurement. *Forests* **4**, 868–886 (2013).

26. Wang, X., Cheng, H., Xi, J., Yang, G. & Zhao, Y. Relationship between park composition, vegetation characteristics and cool island effect. *Sustainability* **10**, 587 (2018).

27. Soga, M., Kanno, N., Yamaura, Y. & Koike, S. Patch size determines the strength of edge effects on carabid beetle assemblages in urban remnant forests. *J. Insect Conserv.* **17**, 421–428 (2013).

28. Lynch, A. J. Creating effective urban greenways and stepping-stones: four critical gaps in habitat connectivity planning research. *J. Plan. Lit.* **34**, 131–155 (2019).

29. Matsuba, M., Nishijima, S. & Katoh, K. Effectiveness of corridor vegetation depends on urbanization tolerance of forest birds in central Tokyo, Japan. *Urban For. Urban Green.* **18**, 173–181 (2016).

30. Cohen, D. A. *et al.* Contribution of public parks to physical activity. *Am. J. Public Health* **97**, 509–514 (2007).

31. Akpinar, A. How is quality of urban green spaces associated with physical activity and health? *Urban For. Urban Green.* **16**, 76–83 (2016).

32. Krekel, C., Kolbe, J. & Wüstemann, H. The Greener, The Happier? The Effects of Urban Green and Abandoned Areas on Residential Well-Being. 65 (2015).

33. Rojas-Rueda, D., Nieuwenhuijsen, M. J., Gascon, M., Perez-Leon, D. & Mudu, P. Green spaces and mortality: a systematic review and meta-analysis of cohort studies. *Lancet Planet. Health* **3**, e469–e477 (2019).

34. Markevych, I. *et al.* Access to urban green spaces and behavioural problems in children: Results from the GINIplus and LISAplus studies. *Environ. Int.* **71**, 29–35 (2014).

35. Chawla, L. Benefits of nature contact for children. *J. Plan. Lit.* **30**, 433–452 (2015).

36. Drinnan, I. N. The search for fragmentation thresholds in a southern Sydney suburb. *Biol. Conserv.* **124**, 339–349 (2005).

37. Lachowycz, K. & Jones, A. P. Does walking explain associations between access to greenspace and lower mortality? *Soc. Sci. Med.* **107**, 9–17 (2014).

38. Wood, L., Hooper, P., Foster, S. & Bull, F. Public green spaces and positive mental health – investigating the relationship between access, quantity and types of parks and mental wellbeing. *Health Place* **48**, 63–71 (2017).

39. Fernández-Juricic, E. Avifaunal Use of Wooded Streets in an Urban Landscape. *Conserv. Biol.* **14**, 513–521 (2000).

40. Sahraoui, Y., Foltête, J.-C. & Clauzel, C. A multi-species approach for assessing the impact of land-cover changes on landscape connectivity. *Landsc. Ecol.* **32**, 1819–1835 (2017).

41. Hamstead, Z. A. *et al.* Geolocated social media as a rapid indicator of park visitation and equitable park access. *Comput. Environ. Urban Syst.* **72**, 38–50 (2018).

42. Pinkham, R. *Daylighting: new life for buried streams*. (Rocky Mountain Institute, 2000).

43. Naiman, R. J., Decamps, H. & Pollock, M. The Role of Riparian Corridors in Maintaining Regional Biodiversity. *Ecol. Appl.* **3**, 209–212 (1993).

44. Fremier, A. K. *et al.* A riparian conservation network for ecological resilience. *Biol. Conserv.* **191**, 29–37 (2015).

45. Flink, C. A. The History of the Rails-to-Trails Movement. *THE DIRT* https://dirt.asla.org/2021/07/06/the-history-of-the-rails-to-trails-movement/ (2021).

46. Rails-to-Trails Conservancy (RTC). Trail-Building Toolbox: Design: Designing for User Type. *Rails-to-Trails Conservancy* http://www.railstotrails.org/build-trails/trail-building-toolbox/design/designing-for-user-type/.

47. Norton, B. A., Evans, K. L. & Warren, P. H. Urban biodiversity and landscape ecology: patterns, processes and planning. *Curr. Landsc. Ecol. Rep.* **1**, 178–192 (2016).

48. Ikin, K. *et al.* Pocket parks in a compact city: how do birds respond to increasing residential density? *Landsc. Ecol.* **28**, 45–56 (2013).

49. Fernández, I. C., Wu, J. & Simonetti Zambelli, J. A. The urban matrix matters: Quantifying the effects of surrounding urban vegetation on natural habitat remnants in Santiago de Chile. *Landsc. Urban Plan.* (2018) doi:10.1016/j.landurbplan.2018.08.027.

50. Goddard, M. A., Dougill, A. J. & Benton, T. G. Scaling up from gardens: biodiversity conservation in urban environments. *Trends Ecol. Evol.* **25**, 90–98 (2010).

51. James, P., Banay, R. F., Hart, J. E. & Laden, F. A review of the health benefits of greenness. *Curr. Epidemiol. Rep.* **2**, 131–142 (2015).

52. Fong, K. C., Hart, J. E. & James, P. A review of epidemiologic studies on greenness and health: updated literature through 2017. *Curr. Environ. Health Rep.* **5**, 77–87 (2018).

53. Alexandri, E. & Jones, P. Temperature decreases in an urban canyon due to green walls and green roofs in diverse climates. *Build. Environ.* **43**, 480–493 (2008).

54. Tietje, W. D., Vreeland, J. K., Siepel, N. R. & Dockter, J. L. Relative Abundance and Habitat Associations of Vertebrates in Oak Woodlands in Coastal-Central California. *Pillsbury Norman H Verner Jared Tietje William Tech. Coord. 1997 Proc. Symp. Oak Woodl. Ecol. Manag. Urban Interface Issues 19–22 March 1996 San Luis Obispo CA Gen Tech Rep PSW-GTR-160 Albany CA Pac. Southwest Res. Stn. For. Serv. US Dep. Agric. P 391-400* **160**, (1997).

55. American Forests. Tree Equity Score Methodology. *Tree Equity Score* https://treeequityscore.org/ (2023).

56. Greco, S. E. & Airola, D. A. The importance of native valley oaks (Quercus lobata) as stopover habitat for migratory songbirds in urban Sacramento, California, USA. *Urban For. Urban Green.* **29**, 303–311 (2018).

57. McKinney, M. L. Urbanization as a major cause of biotic homogenization. *Biol. Conserv.* **127**, 247–260 (2006).

58. Schwartz, M. W., Thorne, J. H. & Viers, J. H. Biotic homogenization of the California flora in urban and urbanizing regions. *Biol. Conserv.* **127**, 282–291 (2006).

59. Belaire, J. A., Whelan, C. J. & Minor, E. S. Having our yards and sharing them too: the collective effects of yards on native bird species in an urban landscape. *Ecol. Appl.* **24**, 2132–2143 (2014).

60. Zuñiga-Palacios, J., Zuria, I., Moreno, C. E., Almazán-Núñez, R. C. & González-Ledesma, M. Can small vacant lots become important reservoirs for birds in urban areas? A case study for a Latin American city. *Urban For. Urban Green.* **47**, 126551 (2020).

61. Anguelovski, I., Connolly, J. J. T., Masip, L. & Pearsall, H. Assessing green gentrification in historically disenfranchised neighborhoods: a longitudinal and spatial analysis of Barcelona. *Urban Geogr.* **39**, 458–491 (2018).

62. Rigolon, A. & Németh, J. "We're not in the business of housing:" Environmental gentrification and the nonprofitization of green infrastructure projects. *Cities* **81**, 71–80 (2018).

03
SITE
STRATEGIES

SITE STRATEGIES

This chapter identifies seven types of urban greenspaces that are major contributors to a city's overall greenspace: Regional and City Parks, Neighborhood Parks, Greenways, Waterfronts, Streetscapes, Schoolyards, and Private Landscapes. Each site type presents its own opportunities to support health and biodiversity based on its spatial configuration, design, *site programming* (or intended uses), and management decisions. Strategies that apply across site types are discussed in Chapter 4, Design Detail and Management Strategies, but are often also referenced in the site-specific sections in this chapter.

Typical contexts and documents in which these strategies may be relevant include:

▶ Community-led visioning for urban greenspaces

▶ Project development and planning (e.g., conceptual and schematic site plans)

▶ Project design (e.g., detailed construction drawings and site plans)

REGIONAL AND CITY PARKS

Regional and city parks are contiguous public greenspaces within the urban fabric that serve either the entirety or significant portions of a city or region. In the context of this guidance, regional and city parks are defined as being at least 5 acres (~2 hectares) in size. For information on parks less than 5 acres, see Neighborhood Parks (pg. 36).

Biodiversity Benefits: Because of their size, large parks can contain large habitat areas that are relatively shielded from urban impacts and support rich mosaics of interconnected habitats that provide varied resources for wildlife, including lakes, wetlands and large forest patches.[1,2]

Human Health Benefits: Large parks have sufficient space to support a wide variety of human uses such as organized sports, walking / running, and immersive contact with nature. They are associated with greater physical, social and psychological health benefits than small parks.[3–5]

KEY TENSIONS

▶ Unregulated human access to park spaces can adversely affect habitat quality for wildlife, especially for urban intolerant wildlife.

▶ Buffer zones that separate high intensity human activities from sensitive habitats are likely to provide limited human functions, but are crucial in maintaining habitat quality for wildlife by reducing anthropogenic and other disturbances.

CONNECTIONS TO PLANNING STRATEGIES

▶ *Large Urban Greenspaces (pg. 22).* Patch size is a key predictor of total species richness,[1,2] and large parks improve the quality of life of city dwellers.[3–5]

▶ *Mitigating Green Gentrification (pg. 28).* Creating large parks could lead to green gentrification in low-income communities. Community engagement is a key strategy for combating gentrification.

SITE STRATEGIES

▶ **Habitat areas.** Regional parks can accommodate habitat types such as lakes, wetlands, and dense forest that are uncommon in other urban greenspaces. A minimum of ~130 acres (~50 hectares) is recommended to create interior habitat that supports urban intolerant wildlife.[2] Where possible and relevant to local ecology, incorporate a variety of habitats interconnected by ecotones to enhance native habitat diversity and connectivity. See Habitat Complexity (pg. 54).

▶ **Gradient of park programming.** Develop a gradient of use intensity for park areas based on adjacent access, land uses, existing topographical characteristics and landscape features. High-use park elements that could disturb wildlife such as sports fields, structures, and lawns that host large events should be sited near park edges to increase public access and reduce disturbance while sensitive habitats should be located near the park center to reduce urban disturbance. Floodable spaces can be included to safely contain potential overflow from nearby waterways.

▶ **Buffers for sensitive habitats.** To reduce disturbance, separate sensitive habitats from park edges and areas of high human use with transitional buffer spaces that host lower intensity use and/or physical barriers, such as topography, a water body, rock outcroppings or dense forest buffer. Buffer widths can be based on the alert distance of species of concern. See Managed Access (pg. 64).

▶ **Circulation design.** Vary the width, length, density and connectivity of the park's pedestrian circulation system to control the intensity of human use in different park spaces. Wide multi-use paths and connected road systems are appropriate for areas designated for higher intensity of human use, while narrow, potentially elevated trails can be used to limit access to some portions of sensitive habitats while providing opportunities for mental restoration.

▶ **Vegetated park edge.** Plant tree buffers and low border vegetation around park edges with a high percent of canopy closure to provide shade in the summer for physical activities, while reducing external disturbances, such as traffic noise and pollution, to provide better foraging and breeding spaces for diverse bird species.[6] Depending on local climate and the shape of a large park, the combination of dense forest edges and large herbaceous patches can support a wide range of physical activities. See Urban Trees and Forests (pg. 58) and Habitat Complexity (pg. 54).

▶ **Nature immersion programs.** Large parks can provide rare opportunities in urban areas for nature immersion programs, such as outdoor education, therapy gardening, and urban forestry management, that require a sizable, programmable greenspace and other facilities. See Garden Spaces (pg. 62).

GREENWAY

THICK FOREST BUFFER
NEAR SENSITIVE HABITATS

LIMITED ACCESS
TO SENSITIVE
HABITATS

ELEVATED
WALKWAY

HIGHLY SENSITIVE HABITAT ZONE

TRANSITIONAL BUFFER SPACES

HIGH-USE PARK ELEMENTS
NEAR THE URBAN CORE

FOREST BUFFER

HIGH DENSITY URBAN CORE

0   100   200   400 ft

NEIGHBORHOOD PARKS

Neighborhood parks are public greenspaces designed to serve a particular neighborhood or community within a city. Such parks are typically less than 5 acres (~2 hectares) and may be too small to include major recreation facilities and large areas of undisturbed habitat. Human uses often dominate these smaller parks. For more information on spaces larger than 5 acres see Regional and City Parks (pg. 34).

Biodiversity Benefits: Small parks provide habitat for native plants and small mobile wildlife such as birds,[7] butterflies, and bees,[8] and act as stepping stones, allowing movement of wildlife through urban spaces.[9,10]

Human Health Benefits: Small parks provide a distributed network of greenspaces that increase accessibility and exposure to recreational, cooling, and other health benefits of greenspace. Having parks close to homes maximizes the many health benefits of greenspace including improved mental health, physical health, and children's behavioral development and connection to nature.[3,11–14]

KEY TENSIONS

▶ Due to space constraints, it is difficult to create undisturbed habitat areas within small neighborhood parks. Small park design may need to prioritize human uses and benefits.

▶ Isolated small patches may act as ecological traps by attracting wildlife but not providing sufficient resources or exposing wildlife to dangers.[15,16]

CONNECTIONS TO PLANNING STRATEGIES

▶ *Community-Driven Projects (pg. 20).* Involving children and the local community in the planning of neighborhood parks allows them to better fit community needs and preferences, which encourages park use[17] and supports children's well-being.[18]

▶ *Greenspace Connectivity (pg. 24).* Having a network of small greenspaces distributed throughout the landscape increases access to parks for humans and connectivity for biodiversity. Planning parks along active transit corridors or greenways and near public transportation further supports park use.[19,20] Ideally small parks should be connected to other greenspaces based on the patch to patch dispersal distance of focal species.

SITE STRATEGIES

▶ **Habitat areas.** Space for undisturbed habitat may be limited in neighborhood parks. Maximize the value of limited area available for habitat by planting native pollinator gardens and key host plants for target species. Design park spaces around high value features such as existing large trees[21] and water features. Constructed habitat features such as bird and bat boxes can augment habitat when space is limited.[22] Wildlife-friendly management practices will maximize benefits of small habitat areas. See Wildlife-Friendly Management (pg. 66).

▶ **Tree Canopy.** Cluster trees to maximize cooling and habitat benefits.[23] Place playgrounds and activity areas under tree canopy to reduce UV exposure and provide shading.[24] Place trees to connect larger patches of habitat for wildlife. See Urban Trees and Forests (pg 58).

▶ **Structural Complexity.** Where possible, introduce multiple layers of vegetation. Structural complexity supports biodiversity in small greenspaces[6,10,25] by providing habitat and food resources for species[26] while providing cooling benefits and supporting children's imaginative play.[18] See Habitat Complexity (pg. 54).

▶ **Separation of Uses.** Separate human use spaces from habitat areas to minimize habitat disturbance.[27] To maximize benefit for people, separate active use spaces such as sports courts and playgrounds and passive use spaces such as picnic tables and benches.[17,28,29]

▶ **Playgrounds and Recreation.** Within human use areas, prioritize recreational areas for children. Children with a nearby park playground are more likely to be classified as healthy weight than children without nearby playgrounds.[28] Including areas for recreational activities such as basketball courts, tennis courts, tracks, and swimming pools boost mental health and physical activity levels.[30,31]

▶ **Water features.** Water features such as ponds and shallow streams can promote social interaction, relaxation, and recreation while providing habitat for biodiversity.[32] Water features can act as "miniature ecosystems" and provide educational opportunities.[33] See Water Features (pg. 60).

▶ **Management for accessibility and activity.** Reduce barriers such as park fees and limited operating hours to increase accessibility and equity for park users.[17] Include programming such as concerts and potlucks to support park use.[17] Maintain facilities to support physical activity and perceived safety.[34,35]

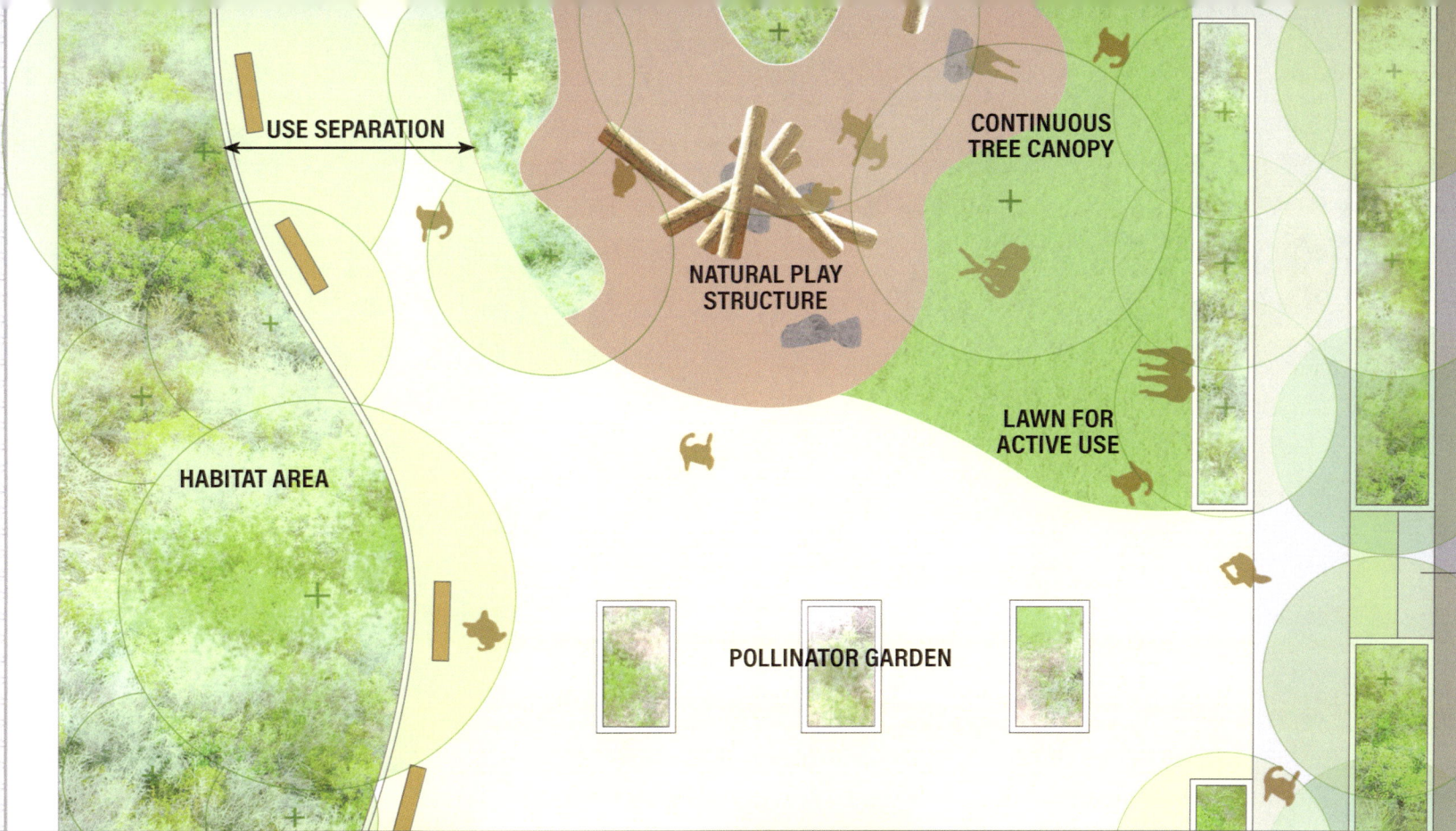

USE SEPARATION

HABITAT AREA

NATURAL PLAY
STRUCTURE

CONTINUOUS
TREE CANOPY

LAWN FOR
ACTIVE USE

POLLINATOR GARDEN

0   10   20   40 ft

GREENWAYS

Greenways are relatively narrow linear greenspaces that usually include multi-use pathways for active transportation (walking, biking, etc.) in addition to vegetated areas. Many greenways are constructed along linear infrastructure such as railroads and utilities that limit development. For more information on greenspaces along water features, see Waterfronts (pg. 40).

Biodiversity Benefits: Greenways allow for movement of wildlife through urban spaces and can form corridors that connect large patches of habitat, particularly important for mammals and birds.[36]

Human Health Benefits: Multi-use paths within greenways create opportunities for active transportation and recreation and are associated with greater outdoor physical activity.[19]

KEY TENSIONS

▶ Human transit and recreation in greenways can limit their habitat benefit, especially if they are heavily used and/or have large areas of hardscape and managed landscape such as lawns.[36,37]

▶ Features helpful for wildlife, such as dense understory vegetation and preservation of dead and downed wood, may signal neglect to the public, reducing recreation value.[36]

▶ In limited space, designers may be forced to choose between recreation and biodiversity goals. Recreation goals are more visible and often take precedence.[36]

CONNECTIONS TO PLANNING STRATEGIES

▶ *Greenspace Connectivity (pg. 24).* Properly sited and designed greenways allow the movement of animals and people through the urban environment and are an important tool in creating a connected network of greenspaces at the city scale.

▶ *Large Urban Greenspaces (pg. 22).* Greenways are most effective at supporting biodiversity when they connect to large greenspaces with interior habitat.[36]

▶ *Green Urban Fabric (pg. 26).* Encouraging adjacent property owners to retain vegetation near greenway and planting trees along adjacent streets can expand greenway functional width for biodiversity support.[36]

SITE STRATEGIES

▶ **Multi-use path located on edge of greenway.** Locate paths on the edge of the greenway to allow for a wider area of undisturbed natural habitat for wildlife, while still facilitating physical activity for people.[36,37] See Additional Resources for sources containing recommendations for path widths.

▶ **Habitat areas with limited public access.** Create wide wildlife movement corridors for urban-intolerant wildlife within the greenway to support a broader array of biodiversity; for example, studies found that bird species richness and nest survival increased in corridors greater than ~160 feet (50 meters) wide.[37,38] See Native Plants (pg. 56), Habitat Complexity (pg. 54), and Managed Access (pg. 64).

▶ **Vegetation structure.** Plant both trees and understory shrubs in habitat areas to increase the variety of available resources for wildlife. More structural complexity and varied vegetation can support more wildlife species.[26] See Habitat Complexity (pg. 54).

▶ **Overlooks and spur trails for habitat viewing.** Create viewing areas to allow for respite and contemplation of sensitive habitat while minimizing human disturbance.[39] See Managed Access (pg. 64).

▶ **Trees lining multi-use path.** Provide tree canopy cover to shade trails, encouraging use in warm weather and reducing UV exposure. Placing trees on the opposite side of natural areas creates a canopy that spans the greenway and can increase the functional size of the habitat area for wildlife that use trees. See Urban Trees and Forests (pg 58).

ADDITIONAL RESOURCES

▶ Creating Effective Urban Greenways and Stepping-stones: Four Critical Gaps in Habitat Connectivity Planning Research. 2018. Lynch, A.

▶ Design Guidelines for Buffers, Corridors, and Greenways. 2008. United States Department of Agriculture.

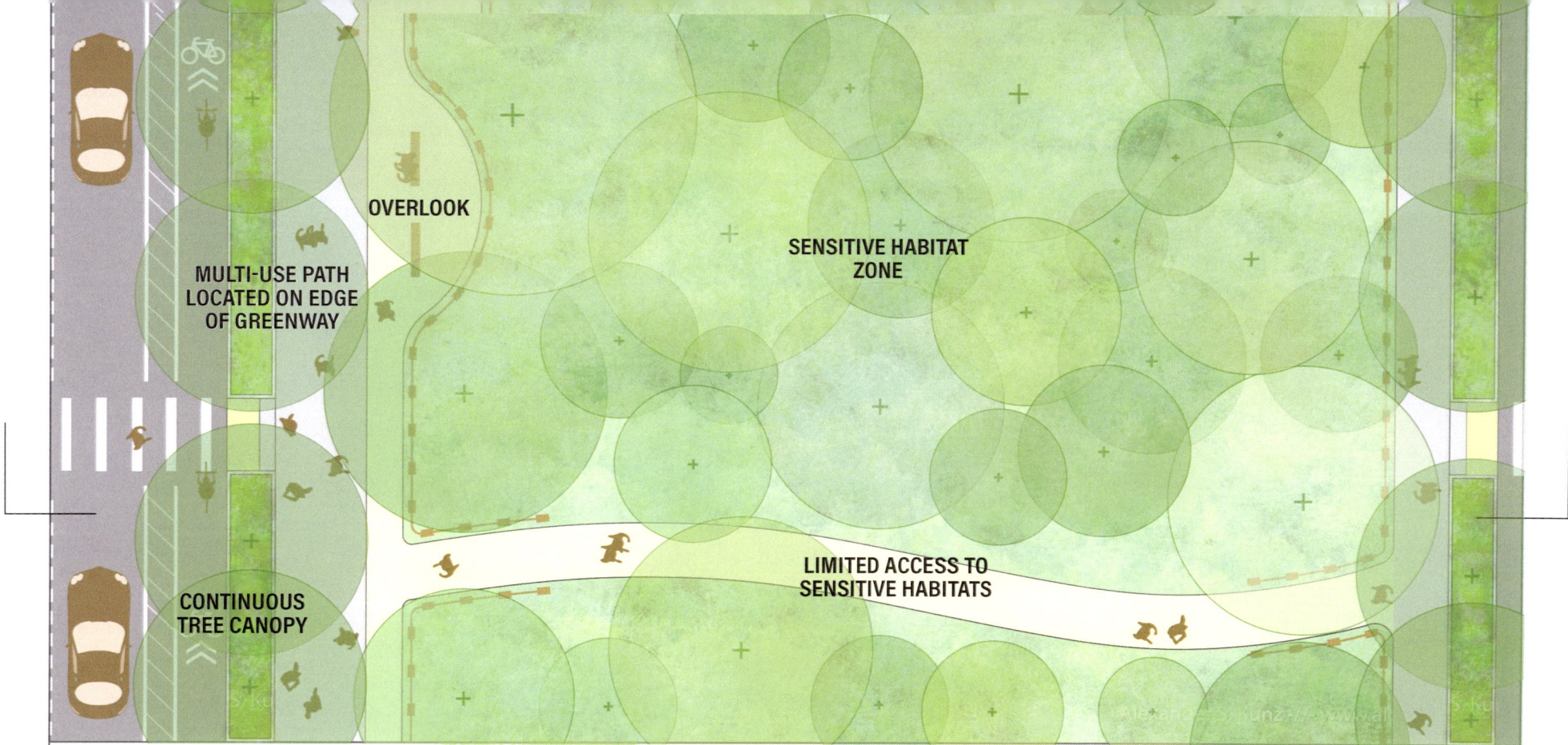

MULTI-USE PATH LOCATED ON EDGE OF GREENWAY

OVERLOOK

SENSITIVE HABITAT ZONE

CONTINUOUS TREE CANOPY

LIMITED ACCESS TO SENSITIVE HABITATS

0   5   10   20 ft

WATERFRONTS

Urban waterfronts are strips of land along a coast, river, or other water body (sometimes referred to as *bluespace*). They vary widely in degree of development and physical form and can include areas of developed high ground, natural or constructed levees, floodplains, riparian forest, and wetlands.

Biodiversity Benefits: Waterfronts often support unique species assemblages due to their proximity to water. They function as ecotones, spanning from terrestrial to aquatic habitats, and also provide corridors for wildlife to move along the shoreline.

Human Health Benefits: Waterfronts can provide recreational opportunities such as walking, swimming, kayaking, and fishing, as well as views of water, which have been associated with improved well-being.[40,41]

KEY TENSIONS

▶ Physical access to water bodies is an important component of recreation in waterfronts. If not properly designed for, human and pet access can damage sensitive habitat features such as wetlands and lead to erosion of streambanks, dunes, and other valuable natural features.

▶ Aquatic habitats are often highly sensitive to urban stressors, so development near these features can limit their ecological value.[42]

▶ High quality riparian habitat often includes dense vegetation that can block the views that would otherwise provide opportunities for contemplation for people and reduce perceived safety of a greenspace.

CONNECTIONS TO PLANNING STRATEGIES

▶ *Greenspace Connectivity (pp. 24).* Waterfronts can provide essential connections to and through adjacent water bodies and habitat patches.

▶ *Mitigating Green Gentrification (pp. 28).* Waterfront redevelopment, especially industrial and/or brownfield, can trigger gentrification.

SITE STRATEGIES

▶ **Visual corridors.** Raised boardwalks and overlooks can be used to provide views for people without impacting riparian vegetation. Seating and picnic tables allow for extended enjoyment.

▶ **Multi-use trails.** Provide multi-use trails to accommodate all non-motorized uses, ages, and abilities to promote higher levels of physical activity and lower levels of cardiometabolic disease.[30] Trails along existing levees can provide views while minimizing the need for new land disturbance. Floodable trails can provide recreation while allowing for natural processes in floodplains.

▶ **Physical access to water.** Direct physical access to the water creates opportunities for swimming, fishing, and boating. Research suggests that connections to the water through recreation and citizen science can improve health and well-being in addition to facilitating greater waterfront stewardship.[43] Creating and managing designated public access points can reduce impact on more sensitive habitat areas.

▶ **Setbacks and wildlife buffers.** Locate trails, pathways, and recreational spaces at a minimum of 200 feet (~60 meters) from marine high tide line and 100 feet (~30 meters) away from other critical or sensitive habitat.[39] Sea level rise threatens to drown many marshes, so plans should include space for marshes to migrate inland over time. Structures, including boardwalks, trails, and pathways should be set back from shorelines, marshes, and other sensitive habitats to reduce impact to wildlife and vulnerability to sea level rise.

▶ **Natural edges and ecotones.** Maintain natural shore edges where possible. Methods that harden shorelines, such as bulkheads and sea walls, can have a significant negative impact on wildlife, including loss of shallow-water and wetland habitat as well as an overall decline in habitat in the immediate area.[44] Natural edges can provide comparable stabilization benefits along with enhanced resilience and ecological function.[45]

▶ **Vegetation structure.** Along many streams and lakes, riparian trees provide important cooling and habitat structure. Shading can help ensure that water temperatures are suitable for aquatic wildlife. See **Native Plants (p 56)**.

▶ **Allow for physical processes**. Re-establish hydrologic and geomorphologic processes (weathering, erosion, and deposition) to boost resilience and long term function. Design features can include sea wall setbacks to allow for beach formation, river-floodplain reconnections to provide space for contained seasonal flooding, and marshes for wave attenuation.

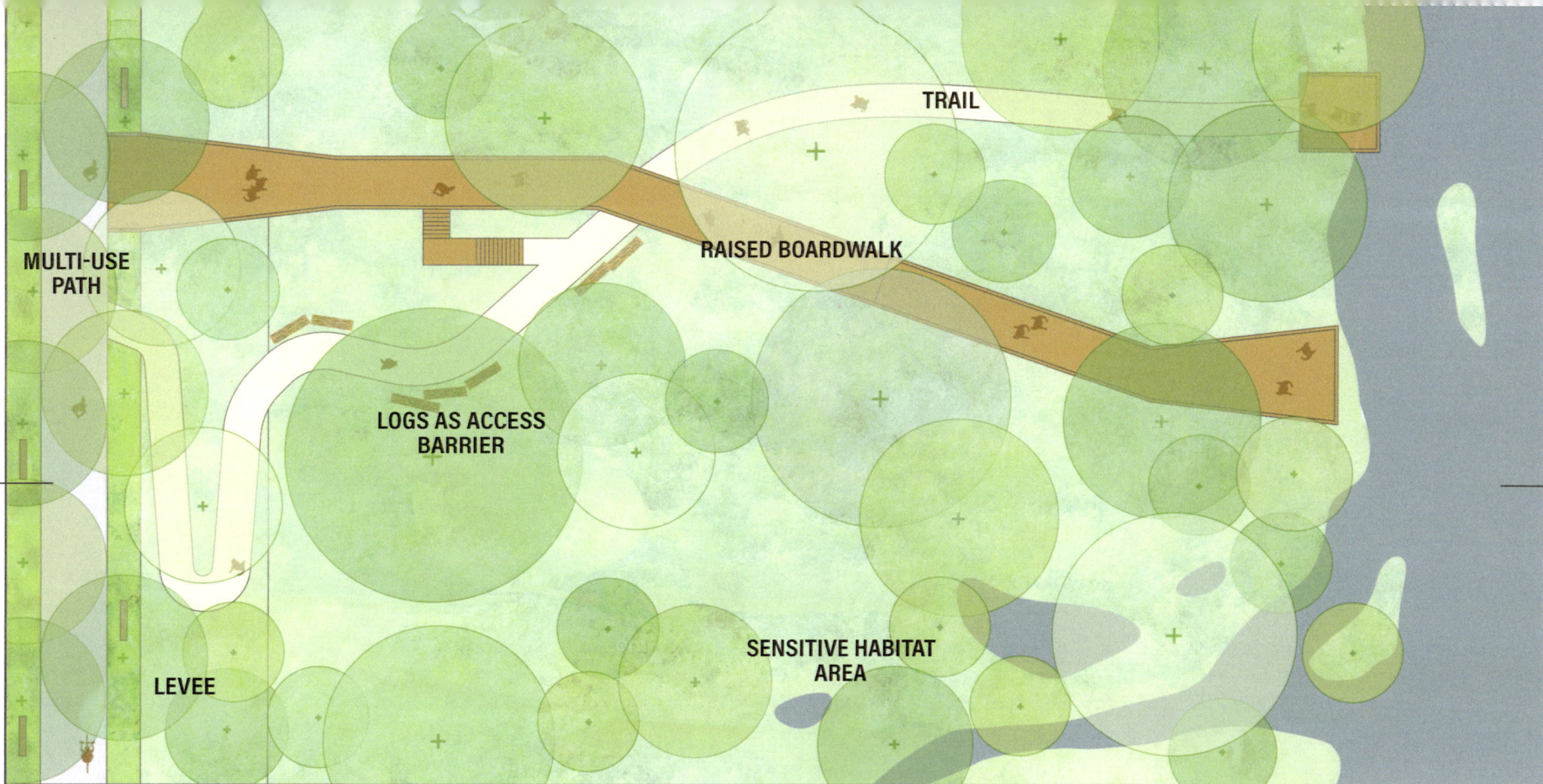

MULTI-USE PATH

TRAIL

RAISED BOARDWALK

LOGS AS ACCESS BARRIER

LEVEE

SENSITIVE HABITAT AREA

0   10   20   40 ft

STREETSCAPES

Streets often make up the largest contiguous public spaces within cities and provide for the movement of both vehicles and people. The streetscape is made up of multiple elements: the roadway, sidewalks, street trees, planters, furnishings (benches, trash cans, street lights, etc.), drainage features and infrastructure for bikes and public transit.

Biodiversity Benefits: Although growing conditions can be some of the harshest in the urban environment, vegetated streetscapes can provide limited habitat for urban-tolerant species and serve as corridors for wildlife.[46,47]

Human Health Benefits: With intentional design, streetscapes can support community health by encouraging walking and biking and by creating public space for social interaction, exercise, and play. Mature tree canopy can provide local relief from sun exposure, extreme heat and air pollution.[48]

KEY TENSIONS

▶ Street trees can conflict with infrastructure, including sidewalks and above ground power lines and can have high maintenance costs. Street trees can drop leaves, fruit, and flowers, and may drop limbs during storms.[49]

▶ Management priorities, such as maintenance of scenic views and space for traffic may constrain management for wildlife, especially on large arterial streets.[50,51]

▶ Native vegetation with high habitat value may not survive the harsh conditions of streetscapes or may have higher maintenance cost than exotic alternatives. Fruiting vegetation or other resources can attract wildlife to roadways, increase the risk of car strikes, and cause maintenance issues.[47]

CONNECTIONS TO PLANNING STRATEGIES

▶ *Green Urban Fabric (pp. 26).* If designed for biodiversity with native plant palettes and appropriate management practices, streetscapes can provide essential connections through urban areas.

SITE STRATEGIES

▶ **Mature street tree canopy.** Place street trees close enough for mature tree canopies to overlap, providing connections through the site for less mobile species. Provide sufficient soil volume to ensure trees can grow to maturity. Achieving 40% tree canopy cover has been shown to increase urban wildlife and reduce impacts from the urban heat island effect.[52]

▶ **Vegetated street verges.** Plant vegetation along linear transportation infrastructure networks to facilitate connectivity for both humans and wildlife.[47,53,54] Design and management decisions that allow visibility for safe crossings should be implemented to prevent vehicles striking pedestrians and wildlife.

▶ **Reduced speed limits on adjacent roadways.** Lower traffic speeds to encourage physical activity and public use.[34] Lower speeds may also increase wildlife movement and reduce car strikes.[55]

▶ **Bulbouts and chicanes.** Install traffic control features, such as bulbouts and chicanes, to reduce vehicle speeds, increase safety for people and wildlife and increase physical activity by children.[34] Well-designed bulbout and chicane interventions can also replace impervious surfaces with potential habitat area.

▶ **Reduced road width.** Reduce the road area on streets and replace with passive use areas for pedestrians such as seating or picnic spaces to provide moments for rest, interaction, and observation. Reduced road width can also reduce vehicle speeds.[56]

▶ **Green stormwater infrastructure.** Incorporate features such as green bioretention cells, flow-through planters, and infiltration basins within streetscapes. Compared to conventional infrastructure equivalents, roadside bioswales can support increased richness and abundance of birds and insects, especially when the habitat is designed for a specific species.[57]

▶ **Overpasses and underpasses.** Create safe roadway crossings for wildlife to avoid car strikes and deaths.[55] A variety of different designs are possible including wildlife and canopy bridges above roads or vegetated culverts, underpasses, and tunnels under roads. Safe crossings are particularly important on roads with high speeds, where animals may not have time to move away from cars. Different wildlife species may need different types of crossings, so crossing designs should be informed by focal species assessments.[47]

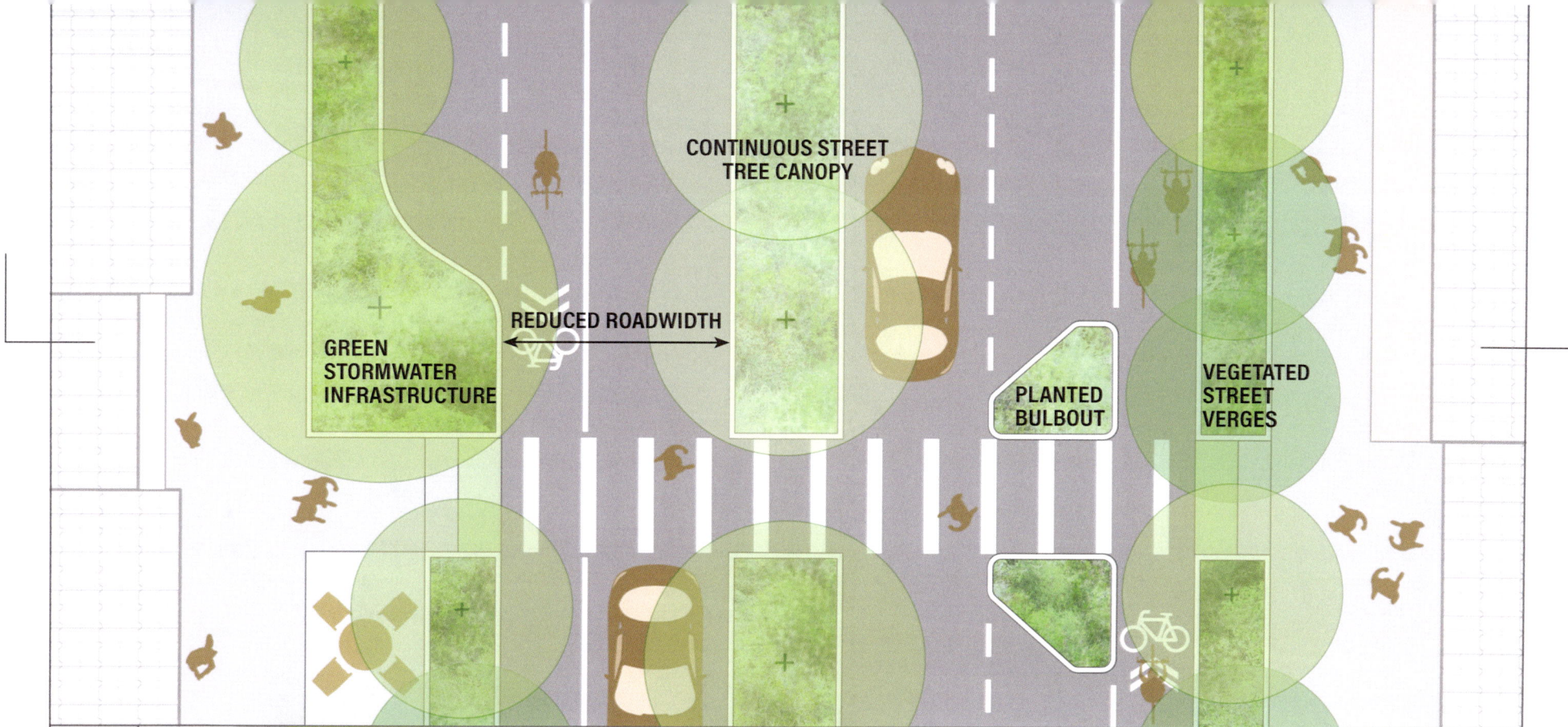

CONTINUOUS STREET
TREE CANOPY

REDUCED ROADWIDTH

GREEN
STORMWATER
INFRASTRUCTURE

PLANTED
BULBOUT

VEGETATED
STREET
VERGES

0   10   20   40 ft

GREEN SCHOOLYARDS

While schoolyards have traditionally been dominated by pavement, green schoolyards aim to create a park-like setting for children to play and learn in. Common features of green schoolyards include trees, water features, nature play areas, and vegetable and native gardens.[58]

Biodiversity Benefits: Green schoolyards can play a similar role as neighborhood parks, providing habitat for small animals and serving as stepping stones within the urban matrix. Green schoolyards also help educate children and parents about native ecosystems and biodiversity.

Human Health Benefits: Maximizing the area of greenspace throughout schoolyards has a multitude of health benefits including improvements in: air quality and overall health,[59] attention and stress recovery in high school students,[60-61] physical activity levels,[62,63] skin microbiota diversity,[64] lower blood pressure,[65] and attention deficit symptom reduction.[66,67] Additionally, greening school yards provides reductions in heat stress and UV exposure.

KEY TENSIONS

▸ Certain aspects of nature play and the potential for children to interact with wildlife may create risks for children. Careful consideration of the types of wildlife that a schoolyard can safely support should inform the planning and design of these spaces. Some level of risk in play spaces can challenge children and create conditions for learning. However, risk should be limited based on the comfort of parents and the surrounding community.[32]

▸ Green schoolyards typically have higher installation and maintenance costs than paved schoolyards[68] and can compete with other educational and recreational priorities for space and financial resources. However, some studies that quantify the economic benefits of green schoolyards indicate that they outweigh the higher costs.[68]

CONNECTIONS TO PLANNING STRATEGIES

▸ *Greenspace Connectivity (pp. 24).* Depending on school location, green schoolyards can increase greenspace access and can be valuable habitat stepping stones.[69] Because schools are distributed relatively evenly throughout cities, green schoolyards can provide benefits such as urban heat island reduction, air pollution mitigation, and off hours greenspace to the local community.

▸ *Green Urban Fabric (p 26).* A green urban fabric surrounding schools supports overall connectivity as well as student's physical activity, as children are more likely to use active travel to commute to school in neighborhoods with high levels of tree canopy and plant diversity.[70]

▸ *Community-Driven Projects (p 20).* Involving children and the local community in the planning of school grounds allows spaces to fit community needs and supports children's well-being.[18] Research has also shown that children are more likely to use and take care of spaces they helped to design or create[71] which in turn supports quality habitat and biodiversity.

SITE STRATEGIES

▸ **Permeable ground cover.** Reduce the use of impervious surfaces and instead use softer natural ground surfaces such as wood chips, mulch, or decomposed granite which can support safe play and habitat connectivity, while reducing heat, erosion, and runoff.[32]

▸ **Classroom views.** Promote views of greenspace from within classrooms (*see image 4*) which has been linked to attention and stress recovery in high school students by bringing the benefits of outdoor greening into the classroom environment.[60,72] Bird blinds and wildlife viewing platforms can increase opportunities for wildlife viewing.[33]

▸ **Tree shading.** Place trees to provide shading (*see image 2*) on playgrounds, gathering spaces, and outdoor classrooms to reduce UV exposure, increase physical activity, and mitigate urban heat island effect.[24,62] Tree canopy coverage is also associated with higher test scores and improved cognitive function.[73]

▸ **Outdoor classrooms and gathering places.** Partially enclosed areas (*see image 1 and 4*) allow children to gather for collective games within natural spaces.[72] Logs, stumps, bamboo, straw, and other natural features can be used to construct outdoor classrooms (*see image 3*) and amphitheaters, while providing habitat.[33,74] Quiet areas can be separated from more active play areas through the use of vegetation, while remaining connected by pathways (*see image 4*).[33] Outdoor instruction is linked to brain development and improved academic performance.[59]

▶ **Edible plants and gardens.** Plant permanently fruiting plants such as blueberries, wild strawberries, blackberries, and raspberries to provide educational opportunities along with food for wildlife.[32] Fruit trees also provide shade and cooling and offer active play opportunities. Gardens provide educational and nutritional benefits for children while attracting multiple species of beneficial insects.[32] See **Garden Spaces (pg. 62).**

▶ **Wildlife habitat.** Create small habitat patches and resources such as pollinator gardens, bird boxes, and water features. Provide different habitat types and structural complexity within habitats (*see image 4*). Tall grasses and bushes help to support imaginative play, while providing habitat for biodiversity.[75] Diverse forested schoolyards have been shown to have an enriching and diversifying effect on children's skin microbiota.[64] Wildlife viewing opportunities such as bird blinds and interpretive signage can support educational benefits of habitat. See **Habitat Complexity (pg. 54)** and **Wildlife-Friendly Management (pg. 66).**

▶ **Nature pieces for play.** Natural loose parts such as downed logs, leaf litter, seeds, rocks, and pinecones can be used for imaginative play while also providing habitat for biodiversity such as worms, toads, chipmunks, and more.[32,74] Avoid clearing leaf litter and debris to maintain nature pieces and habitat. See **Wildlife-Friendly Management (pg. 66).**

❶ Educational area in Brooklyn Botanical Garden's Children's Garden in NY, NY (SFEI) **❷** Shaded swings at Brooklyn Bridge Park in NY, NY (SFEI) **❸** Fiddlehead Forest School in Seattle, WA (SFEI) **❹** School garden at Basisschool de Bijenkorf in Eindhoven, Netherlands (Agaath, CC BY 4.0)

PRIVATE LANDSCAPES

Private residential yards, commercial landscaping, and university and corporate campuses are often dominated by non-native trees, shrubs, and turf. They offer an opportunity to bolster urban habitat through landscape conversion practices.

Biodiversity Benefits: Many private landscapes offer an opportunity to provide habitat on a small scale. These stepping stone interventions repeated at the neighborhood, district, city, and regional scale can help to bolster the overall urban habitat matrix.[76] Private campuses can include large habitat patches.

Human Health Benefits: Native landscaping can provide opportunities for residents and employees to connect with biodiverse greenspaces and observe native wildlife. Both can have a positive impact on mental health, well-being, and sense of place.[17,77,78]

KEY TENSIONS

▶ Private landscapes that attract birds can increase the risk of window strikes[79] and predation by domestic cats.[80]

▶ Turf may encourage human use and match aesthetic preferences but limit opportunities for native plants and wildlife support.

CONNECTIONS TO PLANNING STRATEGIES

▶ *Green Urban Fabric (pp. 26).* Reducing impervious surfaces, vacant lot conversion, and selecting native and wildlife-supporting plants for landscaping all contribute to greening the urban fabric.

SITE STRATEGIES

▶ **Strategically placed uses.** Consider the types of access and views that will be required, and identify areas where native biodiversity support can be prioritized. Many commercial properties contain greenspace that is used minimally if at all, and is often dominated by turf or other low-function cover types with high input and maintenance costs.

▶ **Tree cover.** Identify opportunities on private property to support trees that require more space than street trees. Large and spreading trees, such as oaks, can improve habitat complexity and resource availability. See Urban Trees and Forests (pg. 58).

▶ **Green walls and green roofs**. Install green walls (*see image 4*) and roofs (*see image 2*) with diverse plant assemblages to create additional space for biodiversity support.[81] Compared to conventional building elements, green walls and roofs can support increased richness and abundance of birds and insects.[57] Green roofs that are designed for access by the public or building occupants provide spaces for rest, relaxation, and recovery.[82]

▶ **Fire safety.** In fire prone regions, create defensible space and use fire smart landscaping principles to reduce risk of fire spread to buildings. Using native plants and preventing trees from hanging over roofs are some of several strategies. See Additional Resources section for more information.

▶ **Alternatives to pesticide use**. Reduce pesticide use to limit potential health impacts for both humans and wildlife. Some estimates suggest that homeowners use 10 times more chemical pesticides per acre in urban environments compared to agriculture, leading to a myriad of negative consequences for human health and biodiversity.[83] Consider Integrated Pest Management. See Wildlife-Friendly Management (pg. 66).

▶ **Maintenance.** Develop a maintenance plan for the long term care of commercial properties. More complex and biodiverse plantings may require more or different types of maintenance. See Wildlife-Friendly Management (pg. 66).

▶ **Public education**. Install signs to educate the public about the potential biodiversity support role of private landscapes. Commercial properties may wish to inspire positive feelings in visitors and employees, so clear communication about goals and management may be important. For homeowners, public education may include campaigns to alter ordinances requiring a particular type of landscaping.

▶ **Biodiversity support incentives**. Incentivize biodiversity support on private lands. Aligning individual efforts to provide consistent and complementary biodiversity support across entire neighborhoods can create benefits for a broader suite of species.[76] Existing incentive programs supporting drought tolerant plantings or green infrastructure on private property can be adapted to encourage planting for biodiversity support.

ADDITIONAL RESOURCES

▶ Wildfire Action Plan. 2022. Cal Fire.

▶ Planning, Designing, and Managing Green Roofs and Green Walls for Public Health: An Ecosystem Services Approach. 2022. Sang et al..

1 Greenway at corporate campus in Sunnyvale, CA (SFEI) **2** Green roof at Greenwich Village School in NY, NY (Aloha Jon, CC BY 4.0) **3** Wildlife-friendly backyard garden (Carol Norquist, CC BY 4.0) **4** One Central Park in Sydney, Australia (MDRX, Unsplash)

REFERENCES

1. Nielsen, A. B., van den Bosch, M., Maruthaveeran, S. & van den Bosch, C. K. Species richness in urban parks and its drivers: A review of empirical evidence. *Urban Ecosyst.* **17**, 305–327 (2014).

2. Beninde, J., Veith, M. & Hochkirch, A. Biodiversity in cities needs space: a meta-analysis of factors determining intra-urban biodiversity variation. *Ecol. Lett.* **18**, 581–592 (2015).

3. Akpinar, A. How is quality of urban green spaces associated with physical activity and health? *Urban For. Urban Green.* **16**, 76–83 (2016).

4. Paquet, C. *et al.* Are accessibility and characteristics of public open spaces associated with a better cardiometabolic health? *Landsc. Urban Plan.* **118**, 70–78 (2013).

5. Koohsari, M. J. *et al.* Are public open space attributes associated with walking and depression? *Cities* **74**, 119–125 (2018).

6. Fernández-Juricic, E. Avian spatial segregation at edges and interiors of urban parks in Madrid, Spain. *Biodivers. Conserv.* **10**, 1303–1316 (2001).

7. Huang, Y., Zhao, Y., Li, S. & von Gadow, K. The Effects of habitat area, vegetation structure and insect richness on breeding bird populations in Beijing urban parks. *Urban For. Urban Green.* **14**, 1027–1039 (2015).

8. Osborne, J. L. *et al.* Quantifying and comparing bumblebee nest densities in gardens and countryside habitats. *J. Appl. Ecol.* **45**, 784–792 (2008).

9. Rudd, H., Vala, J. & Schaefer, V. Importance of backyard habitat in a comprehensive biodiversity conservation strategy: a connectivity analysis of urban green spaces. *Restor. Ecol.* **10**, 368–375 (2002).

10. Palmer, G. C., Fitzsimons, J. A., Antos, M. J. & White, J. G. Determinants of native avian richness in suburban remnant vegetation: Implications for conservation planning. *Biol. Conserv.* **141**, 2329–2341 (2008).

11. Rojas-Rueda, D., Nieuwenhuijsen, M. J., Gascon, M., Perez-Leon, D. & Mudu, P. Green spaces and mortality: a systematic review and meta-analysis of cohort studies. *Lancet Planet. Health* **3**, e469–e477 (2019).

12. Richardson, E. A., Pearce, J., Shortt, N. K. & Mitchell, R. The role of public and private natural space in children's social, emotional and behavioural development in Scotland: A longitudinal study. *Environ. Res.* **158**, 729–736 (2017).

13. Zijlema, W. L. *et al.* The relationship between natural outdoor environments and cognitive functioning and its mediators. *Environ. Res.* **155**, 268–275 (2017).

14. Krekel, C., Kolbe, J. & Wüstemann, H. The Greener, The Happier? The Effects of Urban Green and Abandoned Areas on Residential Well-Being. 65 (2015).

15. Battin, J. When Good Animals Love Bad Habitats: Ecological Traps and the Conservation of Animal Populations: *Ecological Traps. Conserv. Biol.* **18**, 1482–1491 (2004).

16. Bonnington, C., Gaston, K. J. & Evans, K. L. Ecological traps and behavioural adjustments of urban songbirds to fine-scale spatial variation in predator activity: Nest predators, territory selection and nest predation rates. *Anim. Conserv.* **18**, 529–538 (2015).

17. Kuo, M. Nature-deficit disorder: Evidence, dosage, and treatment. *J. Policy Res. Tour. Leis. Events* **5**, 172–186 (2013).

18. Chawla, L. Benefits of nature contact for children. *J. Plan. Lit.* **30**, 433–452 (2015).

19. Schipperijn, J., Bentsen, P., Troelsen, J., Toftager, M. & Stigsdotter, U. K. Associations between physical activity and characteristics of urban green space. *Urban For. Urban Green.* **12**, 109–116 (2013).

20. Hamstead, Z. A. *et al.* Geolocated social media as a rapid indicator of park visitation and equitable park access. *Comput. Environ. Urban Syst.* **72**, 38–50 (2018).

21. Greco, S. E. & Airola, D. A. The importance of native valley oaks (Quercus lobata) as stopover habitat for migratory songbirds in urban Sacramento, California, USA. *Urban For. Urban Green.* **29**, 303–311 (2018).

22. Jokimaki, J. Occurrence of breeding bird species in urban parks: Effects of park structure and broad-scale variables. 15 (1999).

23. Tan, Z., Lau, K. K.-L. & Ng, E. Urban tree design approaches for mitigating daytime urban heat island effects in a high-density urban environment. *Energy Build.* **114**, 265–274 (2016).

24. Boldeman, C., Dal, H. & Wester, U. Swedish pre-school children's UVR exposure - a comparison between two outdoor environments. *Photodermatol. Photoimmunol. Photomed.* **20**, 2–8 (2004).

25. Belaire, J. A., Whelan, C. J. & Minor, E. S. Having our yards and sharing them too: the collective effects of yards on native bird species in an urban landscape. *Ecol. Appl.* **24**, 2132–2143 (2014).

26. Matsuba, M., Nishijima, S. & Katoh, K. Effectiveness of corridor vegetation depends on urbanization tolerance of forest birds in central Tokyo, Japan. *Urban For. Urban Green.* **18**, 173–181 (2016).

27. Kuo, M. How might contact with nature promote human health? Promising mechanisms and a possible central pathway. *Front. Psychol.* **6**, 1093 (2015).

28. Kaczynski, A. T., Potwarka, L. R. & Saelens, B. E. Association of Park Size, Distance, and Features With Physical Activity in Neighborhood Parks. *Am. J. Public Health* **98**, 1451–1456 (2008).

29. Baran, P. K. *et al.* Park use among youth and adults: examination of individual, social, and urban form factors. *Environ. Behav.* **46**, 768–800 (2014).

30. Cohen, D. A. *et al.* Contribution of public parks to physical activity. *Am. J. Public Health* **97**, 509–514 (2007).

31. Wood, L., Hooper, P., Foster, S. & Bull, F. Public green spaces and positive mental health – investigating the relationship between access, quantity and types of parks and mental wellbeing. *Health Place* **48**, 63–71 (2017).

32. Moore, R. C. & Cooper, A. *Nature Play & Learning Places*. http://outdoorplaybook. ca/wp-content/uploads/2015/09/Nature-Play-Learning-Places_v1.5_Jan16.pdf (2014).

33. Danks, S. G. *Asphalt to Ecosystems: Design Ideas for Schoolyard Transformation*. (NYU Press, 2010). doi:10.2307/j.ctt21pxmpd.

34. de Vries, S. I., Bakker, I., van Mechelen, W. & Hopman-Rock, M. Determinants of activity-friendly neighborhoods for children: results from the SPACE study. *Am. J. Health Promot. AJHP* **21**, 312–316 (2007).

35. Branas, C. C. *et al.* Citywide cluster randomized trial to restore blighted vacant land and its effects on violence, crime, and fear. *Proc. Natl. Acad. Sci.* **115**, 2946–2951 (2018).

36. Lynch, A. J. Creating effective urban greenways and stepping-stones: four critical gaps in habitat connectivity planning research. *J. Plan. Lit.* **34**, 131–155 (2019).

37. Mason, J., Moorman, C., Hess, G. & Sinclair, K. Designing suburban greenways to provide habitat for forest-breeding birds. *Landsc. Urban Plan.* **80**, 153–164 (2007).

38. King, D. I., Chandler, R. B., Collins, J. M., Petersen, W. R. & Lautzenheiser, T. E. Effects of width, edge and habitat on the abundance and nesting success of scrub–shrub birds in powerline corridors. *Biol. Conserv.* **142**, 2672–2680 (2009).

39. Bentrup, G. Conservation Buffers—Design guidelines for buffers, corridors, and greenways. *Gen Tech Rep SRS–109 Asheville NC US Dep. Agric. For. Serv. South. Res. Stn. 110 P* **109**, (2008).

40. Nutsford, D., Pearson, A. L., Kingham, S. & Reitsma, F. Residential exposure to visible blue space (but not green space) associated with lower psychological distress in a capital city. *Health Place* **39**, 70–78 (2016).

41. Garrett, J. K. *et al.* Urban blue space and health and wellbeing in Hong Kong: Results from a survey of older adults. *Health Place* **55**, 100–110 (2019).

42. Bernhardt, E. S. & Palmer, M. A. Restoring streams in an urbanizing world. *Freshw. Biol.* **52**, 738–751 (2007).

43. Beatley, T. Blue urbanism: Exploring connections between cities and oceans. *Blue Urban. Explor. Connect. Cities Oceans* 1–188 (2014) doi:10.5822/978-1-61091-564-9.

44. Gittman, R. K., Scyphers, S. B., Smith, C. S., Neylan, I. P. & Grabowski, J. H. Ecological Consequences of Shoreline Hardening: A Meta-Analysis. *BioScience* **66**, 763–773 (2016).

45. Smith, C. S. *et al.* Hurricane damage along natural and hardened estuarine shorelines: Using homeowner experiences to promote nature-based coastal protection. *Mar. Policy* **81**, 350–358 (2017).

46. Fernández-Juricic, E. Avifaunal Use of Wooded Streets in an Urban Landscape. *Conserv. Biol.* **14**, 513–521 (2000).

47. Riley, S. P. D., Brown, J. L., Sikich, J. A., Schoonmaker, C. M. & Boydston, E. E. Wildlife Friendly Roads: The Impacts of Roads on Wildlife in Urban Areas and Potential Remedies. in *Urban Wildlife conservation: Theory and Practice* (eds. McCleery, R. A., Moorman, C. E. & Peterson, M. N.) 323–360 (Springer US, 2014). doi:10.1007/978-1-4899-7500-3_15.

48. Baldauf, R. Roadside vegetation design characteristics that can improve local, near-road air quality. *Transp. Res. Part Transp. Environ.* **52**, 354–361 (2017).

49. Koeser, A. K., Hauer, R. J., Miesbauer, J. W. & Peterson, W. Municipal tree risk assessment in the United States: Findings from a comprehensive survey of urban forest management. *Arboric. J.* **38**, 218–229 (2016).

50. Viles, R. L. & Rosier, D. J. How to use roads in the creation of greenways: case studies in three New Zealand landscapes. *Landsc. Urban Plan.* **55**, 15–27 (2001).

51. Carbó-Ramírez, P. & Zuria, I. The value of small urban greenspaces for birds in a Mexican city. *Landsc. Urban Plan.* **100**, 213–222 (2011).

52. Ziter, C. D., Pedersen, E. J., Kucharik, C. J. & Turner, M. G. Scale-dependent interactions between tree canopy cover and impervious surfaces reduce daytime urban heat during summer. *Proc. Natl. Acad. Sci.* **116**, 7575–7580 (2019).

53. Sarkar, C. *et al.* Exploring associations between urban green, street design and walking: Results from the Greater London boroughs. *Landsc. Urban Plan.* **143**, 112–125 (2015).

54. Nawrath, M., Kowarik, I. & Fischer, L. K. The influence of green streets on cycling behavior in European cities. *Landsc. Urban Plan.* **190**, 103598 (2019).

55. Glista, D. J., DeVault, T. L. & DeWoody, J. A. A review of mitigation measures for reducing wildlife mortality on roadways. *Landsc. Urban Plan.* **91**, 1–7 (2009).

56. Farouki, O. T. & Nixon, W. J. The effect of the width of suburban roads on the mean free speed of cars. *Traffic Eng. Control* **17**, (1976).

57. Filazzola, A., Shrestha, N. & MacIvor, J. S. The contribution of constructed green infrastructure to urban biodiversity: A synthesis and meta-analysis. *J. Appl. Ecol.* **56**, 2131–2143 (2019).

58. Children and Nature Network. Green Schoolyards for Healthy Communities. *Children and Nature Network* https://www.childrenandnature.org/schools/greening-schoolyards/.

59. Dadvand, P. *et al.* Green spaces and cognitive development in primary schoolchildren. *Proc. Natl. Acad. Sci.* **112**, 7937–7942 (2015).

60. Li, D. & Sullivan, W. C. Impact of views to school landscapes on recovery from stress and mental fatigue. *Landsc. Urban Plan.* **148**, 149–158 (2016).

61. Taylor, A. F., Kuo, F. E. & Sullivan, W. C. Views of nature and self-discipline: evidence from inntercity children. *J. Environ. Psychol.* **22**, 49–63 (2002).

62. Boldemann, C. *et al.* Impact of preschool environment upon children's physical activity and sun exposure. *Prev. Med.* **42**, 301–308 (2006).

63. Dyment, J. E., Bell, A. C. & Lucas, A. J. The relationship between school ground design and intensity of physical activity. *Child. Geogr.* **7**, 261–276 (2009).

64. Mills, J. G., Selway, C. A., Thomas, T., Weyrich, L. S. & Lowe, A. J. Schoolyard Biodiversity Determines Short-Term Recovery of Disturbed Skin Microbiota in Children. *Microb. Ecol.* (2022) doi:10.1007/s00248-022-02052-2.

65. Xiao, X. *et al.* Greenness around schools associated with lower risk of hypertension among children: findings from the Seven Northeastern Cities Study in China. *Environ. Pollut.* **256**, 113422 (2020).

66. Taylor, A. F., Kuo, F. E. & Sullivan, W. C. Coping with ADD: The surprising connection to green play settings. *Environ. Behav.* **33**, 54–77 (2001).

67. Kuo, F. E. & Faber Taylor, A. A potential natural treatment for attention-deficit/hyperactivity disorder: evidence from a national study. *Am. J. Public Health* **94**, 1580–1586 (2004).

68. Trust for Public Land & MK Think. *California Green Schoolyards: A Cost-Benefit Study.* (2022).

69. Muvengwi, J., Kwenda, A., Mbiba, M. & Mpindu, T. The role of urban schools in biodiversity conservation across an urban landscape. *Urban For. Urban Green.* **43**, 126370 (2019).

70. Larsen, K. *et al.* The influence of the physical environment and sociodemographic characteristics on children's mode of travel to and from school. *Am. J. Public Health* **99**, 520–526 (2009).

71. Chawla, L. & Derr, V. The development of conservation behaviors in childhood and youth. in *The development of conservation behaviors in childhood and youth* 527–555 (Oxford University Press, 2012).

72. Mozaffar, F. & Somayeh Mirmoradi, S. Effective use of nature in educational spaces design. *Organ. Technol. Manag. Constr. Int. J.* **4**, 381–392 (2012).

73. Kweon, B.-S., Ellis, C. D., Lee, J. & Jacobs, K. The link between school environments and student academic performance. *Urban For. Urban Green.* **23**, 35–43 (2017).

74. Spotswood, E. *et al.* Making Nature's City: A Science-Based Framework for Building Urban Biodiversity. *San Franc. Estuary Inst. Publ.* (2019).

75. Derr, V. & Rigolon, A. Participatory schoolyard design for health and well-being: Policies that support play in urban green spaces. *Risk Prot. Provis. Policy* 145–167 (2016).

76. Goddard, M. A., Dougill, A. J. & Benton, T. G. Scaling up from gardens: biodiversity conservation in urban environments. *Trends Ecol. Evol.* **25**, 90–98 (2010).

77. Fuller, R. A., Irvine, K. N., Devine-Wright, P., Warren, P. H. & Gaston, K. J. Psychological benefits of greenspace increase with biodiversity. *Biol. Lett.* **3**, 390–394 (2007).

78. Cameron, R. W. *et al.* Where the wild things are! Do urban green spaces with greater avian biodiversity promote more positive emotions in humans? *Urban Ecosyst.* **23**, 301–317 (2020).

79. Klem Jr, D. Bird: window collisions. *Wilson Bull.* 606–620 (1989).

80. Pavisse, R., Vangeluwe, D. & Clergeau, P. Domestic Cat Predation on Garden Birds: An Analysis from European Ringing Programmes. *Ardea* **107**, 103–109 (2019).

81. Cook-Patton, S. C. & Bauerle, T. L. Potential benefits of plant diversity on vegetated roofs: A literature review. *J. Environ. Manage.* **106**, 85–92 (2012).

82. Sang, Å. O., Thorpert, P. & Fransson, A.-M. Planning, Designing, and Managing Green Roofs and Green Walls for Public Health–An Ecosystem Services Approach. *Urban Ecol. Hum. Health* (2022).

83. Meftaul, I. M., Venkateswarlu, K., Dharmarajan, R., Annamalai, P. & Megharaj, M. Pesticides in the urban environment: A potential threat that knocks at the door. *Sci. Total Environ.* **711**, 134612 (2020).

◄ Apartments in Milan, Italy (Marcus Ganahl, Unsplash)

04

DESIGN DETAIL AND MANAGEMENT STRATEGIES

DESIGN DETAIL AND MANAGEMENT STRATEGIES

This chapter identifies and provides implementation guidance for nine strategies to include throughout the design and management of urban greenspaces. These are non-site specific strategies that apply within each of the site types described in Chapter 3.

Each strategy provides important benefits to human health and urban biodiversity. However, the recommendations in the scientific literature for design elements to support human health are much less specific than the recommendations for biodiversity support. As a result, the implementation guidance for this section tends to place greater emphasis on biodiversity support.

Typical contexts and documents in which these strategies may be relevant include:

▶ Project design and construction documents (e.g., site plans and design details, planting schedules, and specifications)

▶ Design standards and typical details for a park system or other greenspace type

▶ Operations and management plans

Red fox in London, UK (Giedriius, Shutterstock)

HABITAT COMPLEXITY

Habitat complexity can be enhanced at multiple scales in the landscape. Here we focus on three categories: *habitat diversity* includes diversity of habitat types across a landscape, *structural complexity* refers to the variety of vegetation types and structural heights at a site, and *plant diversity* relates to the richness and distribution of species and functional roles in plant communities at that site.

Biodiversity Benefits: Habitat complexity within patches is key to creating quality habitat. Increasing vegetation diversity and complexity enhances species diversity in small greenspaces.[1-3] Different habitats and vegetation types provide different resources, which supports overall biodiversity.[4]

Human Health Benefits: Habitat complexity in urban greenspaces provides many health benefits including supporting well-being,[5,6] inducing positive emotions,[7,8] limiting spread of diseases,[9] and protecting against asthma.[10]

KEY TENSIONS

▶ Habitat complexity may create tradeoffs with perceived safety. Creating openings between patches with clear edges and limiting dense understories is likely to increase a sense of safety for visitors;[11] however, less connected patches with low structural complexity are often unfavorable for biodiversity.

▶ Habitat complexity may create tradeoffs with cultural aesthetic expectations of how a landscape should be designed and maintained.[12]

▶ Lawns and open areas encourage physical activity and social interaction but have low habitat complexity.[13]

IMPLEMENTATION GUIDANCE

▶ **Habitat Diversity.** Include a variety of locally appropriate habitat types such as forests, shrubland, grasslands, lakes, streams, and wetlands throughout the urban landscape. Habitat diversity supports species diversity[3,14] and induces positive emotions, such as happiness, while supporting stress reduction and well-being.[5,7,15] For more information on creating aquatic habitats see **Water Features (pg. 60)**.

- **Create multiple habitat zones.** Plant distinct vegetation communities in zones, using remnant habitats, historical ecology information, and relevant contemporary information and climate change predictions as a guide.[16,17] See **Native Plants (pg. 56)**.

- **Restore and conserve rare habitat types.** Protect patches of remnant and rare habitat types that are uncommon in surrounding landscapes (e.g. sand dunes in San Francisco, CA and montane fynbos in Cape Town, South Africa). Protecting remaining tracts of these habitats and restoring them where they historically existed will benefit the plants and animals that rely on the unique resources they provide.

▶ **Structural Complexity.** Varied vegetation layers allow a single habitat type to provide food, cover, and shelter for a variety of different species (*see images 1-3*). In urban areas structural complexity is associated with higher species richness.[18-20] Varied vegetation structure is tied to stress reduction and improved mood,[21] supports children's adventure play,[22] and is preferred for some recreational settings.[23]

- **Create layers of vegetation that match local ecosystems.** Where appropriate to local ecology (*see image 2*), create layers of vegetation. Landscape plantings can include canopy (large trees), understory (small trees), herbaceous, and groundcover layers. Understory vegetation can be rare in urban environments and its inclusion provides habitat to many species.

- **Incorporate keystone structures.** Keystone structures are locally rare but ecologically important features such as large trees and water bodies.[4]

- **Preserve varied ground cover.** Avoid clearing leaf litter, logs, and stones which provide habitat to many species. See **Wildlife-Friendly Management (pg. 66)**.

- **Separate uses.** Perceived safety concerns and habitat disturbance can be mitigated by separating habitat areas from high traffic human-dominated portions of the landscape. In highly trafficked areas where perception of safety for human users is a concern, create dense vegetation on only one side of the trail/pathway and install vegetation that could conceal a person at least 10-15 feet (~3-5 meters) away from paths.[24]

- **Design and maintain cues to care.** Incorporating elements into the landscape that are recognizable as intentional parts of the design and express that the land is cared for can improve public perceptions of complex greenspace. Use of bold patterns in landscape design, linear planting, flowering plants, framing of habitat, and intentional maintenance of native habitat types/locations (e.g., a mowed strip adjacent to pathways) are all examples of "cues to care."[25]

▶ **Plant Diversity.** Greater native plant diversity supports more species in urban greenspaces.[26] Plant diversity supports psychological well-being,[5] protects against childhood asthma,[10] and is tied to fewer respiratory disease hospital admissions.[27]

- **Provide year-round bloom.** Select a variety of native plants whose bloom times are staggered in order to establish bloom year round. This provides visual interest and support for pollinators, native wildlife, and migratory species throughout the year.

- **Plant a variety of host-specific plant species**. Proactive management of species with known host-specific links to animals can benefit populations of these species. For example, milkweeds support monarch butterflies (*see image 4*) and native oak trees support acorn woodpeckers, cynipid wasps, and other oak specialist species.[28]

- **Use native plants.** See Native Plants (pg. 56).

ADDITIONAL RESOURCES

▶ Planting in a Post-Wild World. 2015. West and Rainer.

▶ Cambridge City Council Biodiversity Toolkit. 2021. Cambridge City Council.

① Sage scrub in Santa Ana, CA (SFEI) ② Desert habitat complexity at Saguaro National Park in Tucson, AZ (Chistoph Von Gellhorn, Unsplash) ③ Washington Park Arboretum in Seattle, WA (Leslie Cross, Unsplash) ④ A monarch on native narrowleaf milkweed in Sunnyvale, CA (SFEI)

NATIVE PLANTS

Native plants have grown in a particular region, ecosystem, or habitat type over thousands of years and have naturally evolved to form symbiotic relationships with native wildlife.[29,30] In North America, native plants are often defined as those present prior to European settlement. They define and structure native habitat types while providing diverse native wildlife support and human health benefits.

Biodiversity Benefits: Native plants themselves are a key component of urban biodiversity and often have specialized relationships with native wildlife.[31,32] Native plants often facilitate wildlife movement through an urban landscape better than exotic plants[33,34] and support a greater abundance and diversity of local pollinators and other insects.[35–37] Native plants are also adapted to the local climate and may be more tolerant of local conditions (including drought).

Human Health Benefits: Native plants contribute to overall biodiversity, which has some direct health benefits. Studies have shown an association between native flowering plants and decreased allergen sensitivity,[38] as well as higher overall plant species richness and improved psychological well-being.[5] Native plants can also contribute to a sense of place.[39]

KEY TENSIONS

▶ Non-native plants may better support human health. For example, in deserts, tree cover can provide protection from heat and encourage more time spent outside, even though trees are not native to the region.[40]

▶ Exotic plantings may be easier to maintain or more compatible with urban infrastructure constraints (e.g., rooting patterns of street trees).

▶ Use of native plants as an alternative to turfgrass and ornamentals may have a higher maintenance cost initially but have a lower long-term maintenance cost.[41]

IMPLEMENTATION GUIDANCE

▶ **Plan for ecological communities.** Native plants evolved to grow in community assemblages that perform complementary functions. Wherever possible, consider planting groups of species that would naturally occur together. See Habitat Complexity (pg. 54) and Wildlife-Friendly Management (pg. 66).

▶ **Use locally adapted seed sources.** Local subpopulations of native species are likely adapted to local conditions.[43] Planting from local seed sources (rather than using cultivars) preserves genetic diversity.

▶ **Select plants based on site constraints.** Urban stressors can limit plant survival. For example, plants must be able to tolerate high temperatures, frequent damage, and, potentially, higher-salinity recycled water for irrigation. Plants used in bioretention features must be able to survive hydrologic extremes of saturated soils and extended dry periods. Urban trees should not destroy sidewalks and property and should provide specific functions (e.g., shading).

▶ **Incorporate keystone plant species.** Native plants that are particularly valuable for native species support are known as keystone species. For example, large native oaks provide food as well as vertical structure and shelter.[44] Identifying and incorporating these species into planting plans can help support diverse insect and wildlife communities.

▶ **Integrate locally rare species.** Managed plantings provide an opportunity to support species that may be locally threatened or rare.[45] Including these rare species in plantings is also a good way to increase overall species richness.

▶ **Evaluate future climate resilience.** Climate change is driving range shifts for many species. For longer-lived species, such as trees, evaluate likely suitability to future climate conditions, including changes in temperature, precipitation, and rising groundwater and increased salinity in coastal areas.

▶ **Include non-native species strategically.** Native species may not always be compatible with site constraints or desired ecosystem services (e.g., shading in desert ecosystems). Where native species are not suitable, consider near-natives or species that are likely to provide needed resources to native wildlife.

OVERSTORY:
Dominant

Morella californica *Salix Lasiolepis*

UNDERSTORY:
Subdominant
*Smaller trees &
tall shrubs*

*Cornus sericea
subsp. occidentalis* *Rubus spectabilis*

UNDERSTORY:
Subdominant
*Low shrubs &
groundcovers*

Carex obnupta *Erythranthe guttata* *Juncus bufonius*

❶ Native flowers in Mountain View, CA (SFEI) ❷ Native plantings based on historical habitats (SFEI) ❸ Gardens in Mountain View, CA (SFEI)

▶ **Source plants grown without harmful insecticides.** Avoid sourcing plants grown with harmful insecticides such as neonicotinoids. Neonicotinoids negatively impact pollinator species[46] and have been associated with adverse developmental and neurological outcomes in humans.[47]

▶ **Develop native plantings based on historical habitat types.** Historical habitat types can provide a guide for the types of ecological communities that may be best suited for a site (*see image 2*). To develop a planting plan based on historical habitats:

- Identify all dominant local historical habitat types. Information about historical habitat distribution, composition, and structure can be synthesized from a variety of archival data sources including maps, photographs, and documents. For information about historical ecology methods, see Grossinger et al.[42]

- Develop illustrations of the structure and composition of all dominant historical habitat types (*see image 2*) based on available historical information.

- Categorize all potential planting areas by size, site function, and local planting conditions, and then evaluate spatial overlap with historical habitat types.

- Develop planting strategies that match each planting area with a historical habitat type. Specify vegetation structure, species composition, ecological value, and human use.

ADDITIONAL RESOURCES

▶ The Historical Ecology Handbook. 2005. Egan and Howell.

▶ The role of 'nativeness' in urban greening to support animal biodiversity. 2021. Berthon et al..

▶ Ecological Horticulture at the Presidio. https://www.sfei.org/projects/ecological-horticulture-presidio.

▶ Messy ecosystems, orderly frames. 1995. Nassauer.

▶ Bringing nature home: how you can sustain wildlife with native plants. 2009. Tallamy.

URBAN TREES AND FORESTS

An *urban forest* is a collection of trees growing within a city. Urban forests can range from remnant native forest patches to trees planted throughout cities, along streets and highways, at commercial properties, and in residential yards. For more information on planning scale guidance for urban trees and forests, see **Green Urban Fabric (pg. 26)**.

Biodiversity Benefits: Trees provide vertical structure, nesting resources, cover, and fruit and floral resources. Large native forest patches function as biodiversity hubs, protecting more urban-intolerant species from urban stressors, and even non-native trees can provide structure or food that benefits native wildlife.[30]

Human Health Benefits: Urban forests improve mood, mental health, immune function, and BMI and can lower prevalence of lung cancer, asthma, heat related mortality, and preterm birth.[48] Well-connected forest patches in proximity to homes (e.g., within ½ mile or 0.8 kilometers) and with good access to other developed areas are associated with lower child BMIs,[49] better health for children,[11] and greater physical activity in adults.[50] Trees also provide shading and cooling, and high tree canopy cover can offset the urban heat island effect.[51]

KEY TENSIONS

▶ Although results are mixed,[48] in some cases increased tree canopy cover has been associated with increased prevalence of asthma and allergens.[52,53]

▶ Tree species preferred for human benefits, such as shading and compatibility with urban infrastructure, may be exotic species that do not provide the same biodiversity benefits as natives.[54]

▶ Increased tree cover has been linked to gentrification, so strong housing policies should accompany greening. See **Mitigating Green Gentrification (pg. 28)**.

IMPLEMENTATION GUIDANCE

▶ **Protect large trees.** Large trees can provide critical support for native wildlife communities,[44] as well as shading and visual interest for people. Parks and larger open spaces offer the most opportunity for large trees.

▶ **Shade strategically.** Shade heavily used areas, including transit stops and active transportation corridors. Plant deciduous trees along the south and west side of buildings to reduce summer heat.

▶ **Evaluate soil quality.** Soil quality, including physical composition, nutrient structure, acidity, level of compaction, and porosity also strongly affects the health of urban forests. Sufficient soil surface exposure in particular is crucial for oxygen exchange, runoff infiltration, and nutrient inputs.[55] Match tree selection to the available soil volume and quality to support healthy trees. Soil compaction in dense urban areas can be mitigated by suspending paving on piers, soil cells, or structural soil.[56]

▶ **Match tree species selection to the site.** Consider desired functions and site characteristics. Use native trees where possible, and target locally appropriate levels of diversity. Evaluate potential disservices, including high pollen production, susceptibility to disease, and maintenance issues. See **Native Plants (pg. 56)**.

▶ **Limit water needs**. In arid regions, weigh the benefits of tree planting against the increased water use. In cities where trees were not present historically, identify near-native and drought-tolerant trees where possible.

▶ **Partner with community groups.** Urban forest projects may contribute to gentrification of neighborhoods. Studies have found that urban trees are associated with increased property value.[57,58] To mitigate potential impacts, it is recommended to partner with local non-profit organizations and government agencies on tree planting efforts.[58] See **Mitigating Green Gentrification (pg. 28)**.

▶ **Encourage forest bathing.** Consider adding features that encourage contemplation and calmness, including small paths, quiet spaces, and benches.

ADDITIONAL RESOURCES

▶ Urban nature for human health and well-being: a research summary for communicating the health benefits of urban trees and green space. 2018. United State Department of Agriculture.

▶ Re-oaking. https://www.sfei.org/projects/re-oaking.

1 Urban forest in Portland, OR (Cristofer Maximilian, Unsplash) 2 Trees in Berkeley, CA (SFEI) 3 Trees along HWY 285 in Atlanta, GA (Samuel Agbetunsin, Unsplash) 4 Trees in Brooklyn, NY (SFEI)

WATER FEATURES

Small urban water features such as ponds, wetlands, small fountains, and pools can be components of many urban greenspaces. When properly designed, water features can provide a unique amenity for both people and wildlife. For larger features such as lakes, streams, rivers, and coastlines, see **Waterfronts (pg. 40)**.

Biodiversity Benefits: Impacts to hydrology and water quality limit the biodiversity of most urban water features,[59,60] but these features can still support many wildlife species,[61-63] and provide an important source of freshwater for all wildlife in the landscape. Even fountains can support invertebrate communities[64] and provide an important resource for birds.[65]

Human Health Benefits: Water features have numerous health benefits including supporting park visitation,[66] cooling,[67] physical activity,[68,69] and mental health.[69] Water features are associated with bonding to place and support positive emotion and mental restoration.[70]

KEY TENSIONS

▶ Stationary urban water bodies have the potential to promote disease vectors such as mosquitos as well as toxin-producing algae (cyanobacteria) blooms which may pose a human health hazard.[60]

▶ Some elements of ornamental water features that create human enjoyment may not support biodiversity goals. Water features free of dense vegetation are unlikely to support high biodiversity and steep, formal edges reduce access for wildlife.[63] Ornamental plants and fishes may compete with native species.[60,62]

▶ Water features at sites with the potential for the greatest human benefit (e.g., near busy areas, along major transit corridors) are likely to have high levels of pollution and lower habitat quality. These water features often have impaired water quality and may function as ecological traps.[71]

IMPLEMENTATION GUIDANCE

▶ **Create large water features.** Larger water features support more species; where larger features are not possible, include small features that can act as stepping stones and breeding sites.[60]

▶ **Diversify features.** Where possible include a diversity of water body types (*see image 3*) (varying in size, water chemistry, hydroperiod, stationary/flowing, fish/fishless) as well as structural habitat diversity within the water feature itself to support a wider variety of wildlife species.[60,63] Absence of fish is linked with higher invertebrate and amphibian diversity.[62]

▶ **Support managed access.** Place water features near walkways or other high use areas to improve human benefits. Design for visual access can provide benefits while limiting human disturbance of plants and wildlife. See **Managed Access (pg. 64)**.

▶ **Include aquatic vegetation.** Include submerged, floating, and emergent vegetation (*see image 2*) as locally appropriate to support a diverse community of amphibians, fish, birds, and invertebrates.[62,63]

▶ **Use natural structure.** Natural and spatially complex bed materials with sediment, cracks, and crevices (*see image 1*) support greater biodiversity. Shallow vegetated shores improve habitat quality.[60,63] Vertical walls and concrete that prevent wildlife access should be avoided.[62]

▶ **Manage heat.** Trees placed to shade water features (*see image 2*) reduce daytime temperatures and promote human enjoyment.[60] Shallow vegetated shores also promote cooling.[67] See **Waterfronts (pg. 40)**.

▶ **Use ecological pest management**. Higher species richness, particularly of insects, can support more predators that control mosquito populations and limit the spread of disease.[63,72]

▶ **Provide islands and rocks for basking.** Water features with structural complexity, including rocks that reach the surface, floating debris, and islands, provide safe locations for wildlife that spend time on land. These features also create opportunities for people to view wildlife without disturbing them.

▶ **Daylight streams.** Daylighting piped and culverted waterways is an excellent method for adding water features to an urban environment.[73] See **Greenspace Connectivity (pg. 24)**.

ADDITIONAL RESOURCES

▶ Anthropogenic refuges for freshwater biodiversity: Their ecological characteristics and management. 2013. Chester and Robson.

▶ Daylighting: new life for buried streams. 2000. Pinkham.

▶ Pond Design Principles for Biodiversity. 2013. Fresh Water Habitats.

1 Lincoln Park South Pond in Chicago (Ranjit Souri, Unsplash) **2** Small pond in Brooklyn Bridge Park in NY, NY (SFEI) **3** Diversity of water features within Golden Gate Park in San Francisco, CA (Google Earth)

GARDEN SPACES

Urban garden spaces can come in the form of community farms, rooftop gardens, backyard plots, and edible landscaping. In this section we focus on garden spaces for cultivation. For residential yards see **Private Landscapes (pg. 46)**. Garden spaces provide therapeutic and community benefits and can be designed to maximize biodiversity support. Garden spaces can be incorporated into various site types to promote their human health and biodiversity benefits.

Biodiversity Benefits: Gardens can incorporate diverse native plant assemblages. Including locally rare or threatened native species in gardening spaces can also improve local soil fauna diversity and fauna biomass.[74] Urban gardens provide pollination, seed dispersal, and pest management services to neighboring landscapes[75] and support birds,[76] butterflies,[77] arthropods, spiders, grasshoppers, bees, and beetles.[78] In addition, wildlife-friendly gardens can improve habitat quality and connectivity in the urban fabric.[79]

Human Health Benefits: Garden spaces provide opportunities for gardening, which has important benefits to mental health, stress reduction, well-being, physical activity levels, general health, and BMI while supporting health benefits of access to fresh produce.[80] Community-based gardening such as planting in parks and urban restoration areas, provides opportunities for social interactions and increased exposure to nature and sunlight outdoors.[81] Past studies have also revealed the antidepressant effects of gardening,[82,83] including on individuals with disabilities.[84]

KEY TENSIONS

▶ Many people have a cultural preference for highly manicured gardens dominated by exotic annuals, and diverse gardens with native plants can be viewed as less beautiful and inviting.

▶ Wildlife are often seen as pests within urban gardens. While beneficial insects are welcome, barriers to safely separate wildlife-dedicated space from productive areas such as insect barriers, bird protective fencing, and root protection may be needed.

IMPLEMENTATION GUIDANCE

▶ **Make gardening spaces inclusive.** Design for a broad spectrum of users, including the disabled and the elderly, if the gardening space is intended for public use. Components to consider per the Americans with Disabilities Act (ADA):

- Access such as ramps and handrails on accessible routes

- Height of raised planters accessible to individuals in a wheelchair and seniors with limited mobility

- Resting areas and shelter

▶ **Encourage multiple sensory experiences.** Therapeutic gardening programs that engage with multiple sensory experiences, such as smell, touch, sound and sight, help reduce stress and improve mood.[86,87]

- Consider using host plants as well as water features that attract singing birds and insects that produce soothing natural sounds.

- Planting fragrant, flowering plants provide food sources for wildlife year round while enhancing the sensory experience of gardening.

▶ **Mitigate soil contamination.** Urban soils often contain toxic trace metals such as lead which can be transferred through consumption of soil or contaminated produce.[88] Using raised planting beds as well as considering nearby pollution sources before placing gardens can reduce trace metal risks.[89] Placing edible garden beds at least 30 feet (10 meters) from the road or separating beds from the road with woody vegetation can reduce pollution deposition from cars.[89]

▶ **Create forest gardens.** Designing polyculture forest gardens with tree overstory, herbaceous middle story, and an understory of vegetation, herbs, and flowers can support biodiversity while providing food.[89] See **Urban Trees and Forests (pg. 58)**.

▶ **Integrate structural complexity.** Despite many garden plants being exotic species, structural complexity (*see image 4*) can still support vertebrate diversity through provision of cover, breeding sites, and shelter.[78,90] See **Habitat Complexity (pg. 54)**.

▶ **Include edible landscaping.** Fruit and seed bearing trees and plants can produce food for species while providing foraging opportunities for urban residents to reduce hunger and support social interactions.[91] Trees can be foraged for food, medicine, and resources.

▶ **Plant native flowers.** Including a variety of native floral resources within gardens increases plant diversity and helps to support arthropods, spiders, grasshoppers, bees, beetles, and butterflies.[77,78] See **Native Plants (pg. 56)**.

▶ **Reduce chemical inputs**. To prevent negative impacts of intensive agricultural management, utilize organic and integrated pest management practices within gardens (*see image 1*). See **Wildlife-Friendly Management (pg. 66)**.

ADDITIONAL RESOURCES

▶ Universal Design: Gardens. https://www.asla.org/universalgardens.aspx.

▶ What's a Forest Garden Anyway? Here's How to Create One in Your Backyard. https://www.stateforesters.org/2022/03/30/whats-a-forest-garden-anyway-heres-how-to-create-one-in-your-backyard/.

▶ Gaia's Garden: A Guide to Home-Scale Permaculture. 2009. Toby Hemenway.

① Mulching at Hayes Valley Urban Farm in San Francisco, CA (Chris Martin, CC BY 4.0) **②** Alemany Farm in San Francisco, CA (SFEI) **③** Sartain Street Community Garden in Philadelphia, PA (SFEI) **④** Washington Park Garden in NY, NY (SFEI)

MANAGED ACCESS

Urban greenspaces must manage human–wildlife interactions and human and pet disturbance. This section highlights strategies for controlling access to sensitive habitat, while maintaining views and awe-inspiring greenspace experiences.

Biodiversity Benefits: Limiting access to sensitive habitat reduces human and dog disturbance of plants and wildlife.[92,93] Reduced disturbance can enhance habitat value and allow more species to readily use these spaces.

Human Health Benefits: Managed access can reduce human-animal conflicts[94] and support visitor enjoyment and experience of awe without impacting habitat quality.[95,96]

KEY TENSIONS

▶ Access restrictions for habitat protection can create controversy.[97] Park closures or access limitations may disproportionately impact historically disinvested neighborhoods, due to their lower overall greenspace access.[98]

▶ Poorly placed fences can interfere with wildlife movement and create barriers.

▶ Habitat quality can be negatively impacted in areas of concentrated human use, or where humans are allowed to move freely

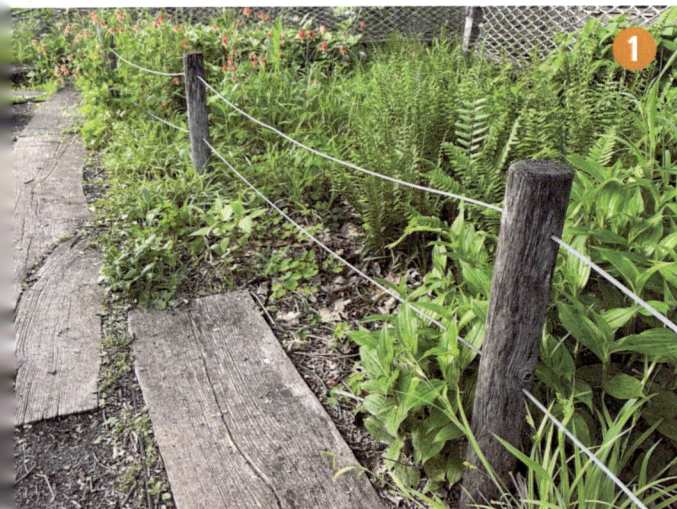

IMPLEMENTATION GUIDANCE

▶ **Create greenspace zones.** Design distinct zones based on human use and habitat sensitivity.[99] Highly sensitive areas, such as breeding habitat, can have restricted access. Moderate sensitivity areas can allow limited access or limit certain activities, such as dog walking. High use areas such as paved pathways or recreation spaces can be sited away from sensitive habitat.

▶ **Incorporate natural barriers.** Fences can block animal movement and make users feel as if they are not in a natural environment, reducing the awe-inspiring experience of nature.[96] Use streams, ditches, steep grades, and screening vegetation (*see image 2*) such as thorny or large bushes as natural barriers to dissuade human access while allowing animal movement.[99]

▶ **Minimize fencing and other barriers to connectivity.** Minimize barriers to allow wildlife to access greenspaces and move through the urban landscape more freely.

▶ **Use wildlife-friendly fencing.** In locations where fencing is needed, use wood and rope fencing <4 feet (~1 meter) high with a 4-6 inch (~10-15 centimeter) gap along the bottom (*see image 1*). These types of low height wildlife-friendly fences allow for animal movement,[100] avoid blocking views,[100] and have been found to be successful at limiting human and dog disturbance to habitat.[93] In aquatic settings, buoys and ropes can be used to designate sensitive habitat areas.

▶ **Make paths well-defined and interesting.** Create paths that offer varied and interesting experiences[95] while discouraging users from leaving designated areas and paths on informal trails.[100] Avoid placement of trails through sensitive areas.[99]

▶ **Place benches strategically.** Avoid placing benches in areas with high habitat value to lower disturbance. Benches and tables should be directed inwards to avoid facing city noise,[101] or situated to provide views (*see image 3*).

▶ **Design entry points far from sensitive habitat.** Human use decreases with distance from entry points,[100] so place parking and staging areas away from sensitive habitat to reduce public access to these areas.

▶ **Use lighting to guide use.** Limit lighting in sensitive habitat zones to discourage use at night[92] while lowering lighting impact on wildlife. See **Limited Outdoor Lighting (pg. 68).**

▶ **Create areas for viewing scenery and wildlife.** Observation points create predictability of human presence for wildlife, while optimizing views for visitors.[100] Often, wildlife disturbance is associated with users' desire to take pictures;[96] supporting managed access to views and photo opportunities (*see image 3*) discourages users from leaving designated areas and disturbing wildlife.

▶ **Provide educational signage for guided access.** Educational signage can educate users on the location and value of habitat as well as consequences of user's actions. Signage can vary from labeling plant species to language such as "no access protected wildlife area" (*see image 4*). See **Wildlife-Friendly Management (pg. 66)**.

ADDITIONAL RESOURCES

▶ Fencing with Wildlife in Mind. 2009. Colorado Parks and Wildlife.

❶ Wildlife-friendly fencing at Brooklyn Botanical Gardens in NY, NY (SFEI) ❷ Walking path in Sunnyvale, CA (SFEI) ❸ Bench at Stow Lake in San Francisco, CA (SFEI) ❹ Signage limiting access at High Line Park, NY, NY (SFEI)

WILDLIFE-FRIENDLY MANAGEMENT

Wildlife-friendly management includes a variety of landscape management practices that seek to reduce human impacts to improve habitat quality by creating more natural conditions. Practices can include reducing chemical inputs and minimizing vegetation maintenance to create greater habitat complexity.

Biodiversity Benefits: Wildlife-friendly management practices improve habitat quality by increasing habitat complexity and minimizing human disturbance and toxic exposure for wildlife. These practices have been tied directly to increases in biodiversity.[41,102]

Human Health Benefits: Wildlife-friendly management strategies can lead to improved air quality and reduced exposure to toxic substances. In particular, wildlife-friendly strategies include the reduction of chemical inputs such as fertilizers, pesticides, and herbicides, which contribute to cancer, birth defects, and other negative health impacts.[103] Also, fewer emissions and less airborne debris from mowers and leaf blowers improves air quality.[104]

KEY TENSIONS

▶ Cultural preferences for highly manicured greenspaces[105] may require more intensive practices, including mowing, pruning, leaf blowing, and herbicide use, that reduce value for wildlife.

▶ While pruning and removal of dead trees and logs removes habitat for wildlife, it may be desired for safety and aesthetic reasons. Selective pruning can allow for risk mitigation while still preserving habitat and lowering maintenance costs for cities.[106]

▶ Fire-prone regions must balance landscaping for biodiversity with fire risk. Near structures, defensible space and fire smart landscaping principles may require clearing of tree limbs, dead trees, leaf litter, and other materials beneficial to biodiversity. See Additional Resources.

IMPLEMENTATION GUIDANCE

▶ **Implement Integrated Pest Management.** Integrated Pest Management (IPM) involves a multi-tactic coordinated and adaptive approach to controlling pests including insects, pathogens, weeds, and rodents that is ecologically supportive and reduces chemical inputs.[107] Examples of IPM include the use of mulch, biological control through beneficial insects and competitive plants, mechanical control such as rat traps, and minimized chemical control such as using target pesticides minimally or use of bait traps as opposed to insecticide sprays.

▶ **Plant native pollinator hosts.** Plant native nectar-producing species that attract native insects to support biological control of unwanted pests, reducing the need for pesticides harmful to human health. Where possible, assemble plant species to create year round bloom for native pollinators. See Habitat Complexity (pg. 54).

▶ **Minimize vegetation maintenance.** Avoid pruning and clearing leaf litter and debris to maintain habitat quality (*see image 2*). Dead snags, downed logs, leaf litter, and other organic material can provide resources and habitat for wildlife such as worms, toads, lizards, birds, and small mammals and can be used for children's imaginative play.[22,108] Reducing maintenance also limits airborne particulate pollution from gas-powered equipment. During the plant installation phase, accommodate for the plant's mature size and leave sufficient space between the planting location and the edge of the planting bed to reduce pruning needs after establishment.

▶ **Build structure for wildlife.** Add features such as bird houses, bee hotels (*see image 1*), and bat boxes to the landscape to help support native species. Incorporate patches of bare ground free of mulch to provide habitat for ground nesting bees and other beneficial insects who prefer bare soil.[116] Sand piles can be added to provide nesting sites for bees.

▶ **Schedule maintenance strategically.** Landscape maintenance should be scheduled strategically to minimize impacts to wildlife. For example, pruning or mowing should occur outside of the breeding season for local wildlife, to avoid any disturbance of nests or juveniles. Mowing should also be timed to allow seed set of native annuals. Landscaping specifically designated to provide habitat, such as pollinator gardens, should minimize maintenance disturbance as much as possible.

▶ **Include educational opportunities.** Interpretive signage (*see image 3*) and art, such as sculptures and murals, can be used as an opportunity to educate the public about the ecological benefits of management practices that may otherwise be interpreted as messy or a sign of neglect.

- **Reduce mowing.** Intensive lawn mowing practices can decrease plant and invertebrate diversity, while increasing pests.[109] Reducing mowing frequency to just a few times a year or shifting to partial mowing increases plant and beneficial insect diversity.[102,110] Reducing mowing frequency also reduces toxic emissions from landscape equipment.[104] In amenity planting areas where 'clean' edges are needed to maintain certain visual character and geometry, consider mowing only the edges, allowing denser inner planting for wildlife shelter.[25]

- **Reduce chemical inputs.** Chemical inputs, including fertilizers, pesticides, and herbicides have profound impacts on biodiversity and human health. Glyphosate, the most widely used herbicide, has been linked to cancer, shifts in microbial diversity, a rise in plant and animal pathogens, and antibiotic resistance.[111] Pesticides increase risk of cancer, birth defects, and preterm birth.[103,112,113] Chemical inputs negatively impact biodiversity, and can have both target and non-target effects on food webs.[114,115] Alternatives, including switching to a lower maintenance landscaping design or IPM, can improve both biodiversity and human health outcomes.

ADDITIONAL RESOURCES

- Introduction to Integrated Pest Management. 2012. Flint and Bosch.

- Cambridge City Council Biodiversity Toolkit. 2021. Cambridge City Council.

- Nests for Native Bees. https://www.xerces.org/publications/fact-sheets/nests-for-native-bees.

- Building and Managing Bee Hotels for Wild Bees 2017. Isaacs

- Gardening for Pollinators. https://www.fs.usda.gov/managing-land/wildflowers/pollinators/gardening#:~:text=Use%20a%20wide%20variety%20of,climate%2C%20soil%20and%20native%20pollinators.

- Bat Gardens and Houses. https://www.batcon.org/about-bats/bat-gardens-houses/.

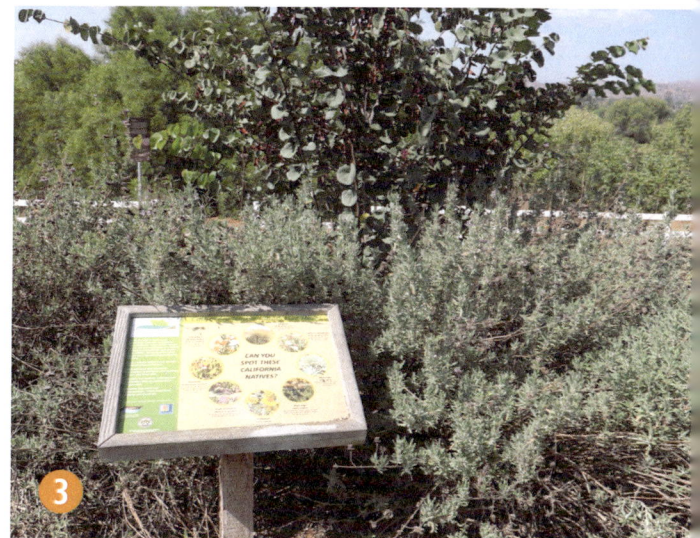

1 Native bee home (Jacopo Werther, Unsplash) **2** Undisturbed leaves and logs within picnic area (Sheila Brown, Unsplash) **3** California native plant interpretive signage along Santa Ana River in CA (SFEI)

LIMITED OUTDOOR LIGHTING

Due to the negative impact of lighting on wildlife and human health, practices limiting the amount, intensity, and high color temperature of lighting can have positive impacts on wildlife and humans.

Biodiversity Benefits: Limiting artificial light at night mitigates negative impacts on wildlife communication, orientation, reproduction timing, predation, habitat selection, circadian rhythm, plant phenology, and ecosystem services.[117]

Human Health Benefits: Artificial lighting is linked to increased breast and prostate cancer risk, increased cortisol, increased vector borne disease risk, and disruptions to circadian rhythm and melatonin production.[118,119] Disruptions to circadian rhythm are associated with negative impacts to psychological, cardiovascular, and metabolic functions.[119] Limiting light mitigates these impacts.

KEY TENSIONS

▶ While reduced lighting itself may not lead to increased crime,[120] it is tied to reductions in perceived safety.[121] Perceived safety is important for the equitable use of public space, and decreased perception of safety due to low lighting can lower greenspace usage.[122]

▶ Outdoor lighting has been linked to physical activity in parks and sports facilities and is important for utilization of a space after dark.[68,123] Adequate lighting influences how people use public spaces at night, with higher use in well lit areas.[124] Lighting can be harnessed to direct users to certain well lit high use areas of public greenspace, while limiting use of areas kept dark for habitat.

IMPLEMENTATION GUIDANCE

▶ **Limit amount of lighting.** The most effective way to lower the impact of artificial lighting is to reduce the amount of overall light.[125] Lighting can be limited to high use areas, leaving greenspace for habitat unlit. Lighting timers and motion sensors can be used to turn lights off when they are not in use. Lighting curfews can be used to limit lighting to high use times.

▶ **Reduce light trespass.** Many light fixtures cause light to shine into areas where it is not needed, such as the sky or sensitive habitat (*see image 1*). In addition, uplighting and light trespass are major contributors to skyglow which negatively impacts wildlife and humans.[117,125,126] Shielded fixtures (*see images 2-5*) can direct light to where it is needed and prevent trespass into areas it is not. The backlight, uplight and glare (BUG) system developed by the International Dark-Sky Association can be used to choose low BUG rating fixtures that reduce light trespass and glare.

▶ **Limit lighting intensity.** Lower the intensity of artificial lighting to reduce negative impacts (*see image 5*). Intensity of lighting has been tied to breast cancer risk and melatonin suppression.[119] Setting a maximum illuminance of 1-3 lux in sensitive habitat areas can reduce impact to species, while providing enough light for people to see.[127]

▶ **Avoid blue-white light.** While species differ on the colors of light they are most sensitive to,[128] broad spectrum blue lighting (>3000 Kelvin) is the most harmful to both wildlife and human health.[119,125,129] Using narrow spectrum lighting, with warmer color temperatures <2700 Kelvin (*see image 4*) can reduce negative impacts to wildlife and human health.[125,129,130]

▶ **Consider target species.** Impacts of specific lighting practices differ by species. While reducing the overall amount of lighting in an area will have the broadest benefits for all species, certain interventions may be more beneficial to specific species. For instance, yellow low pressure sodium lighting can be used to mitigate impacts to sea turtles[131] while red lighting was found to be the least disruptive for migrating birds.[132] Species of concern should be considered when implementing lighting mitigation strategies.

▶ **Use lighting based on site context.** A wetland area along a suburban greenway will have different lighting needs than a green street within the urban core. The Model Lighting Ordinance (MLO) developed by the International Dark Sky Association uses lighting zones to provide lighting guidance based on site context.[133]

IDEAL WILDLIFE-FRIENDLY LIGHTING ④

ADDITIONAL RESOUCES

▶ Model Lighting Ordinance. https://www.darksky.org/our-work/lighting/public-policy/mlo/.

① Lighting that trespasses into sky and surrounding habitat ② Shielded lighting directed onto path ③ Shielded lighting with lower light intensity ④ Ideal wildlife-friendly lighting with shielded, lower light intensity, and warmer color temperature light ⑤ Shielded lighting in NY, NY (Koushalya Karthikeyan, Unsplash)

ALTERNATIVE GROUNDCOVERS

Turfgrass lawn monocultures dominate the urban landscape and require high levels of chemical inputs, irrigation, and management. Urban greenspaces dominated by lawn monocultures have low biodiversity and the intensive maintenance they require creates negative impacts on human and ecosystem health. Many lawns can be replaced with meadows, grass polycultures, shrubs, mulch, or other ground cover alternatives.

Biodiversity Benefits: Most alternatives to lawn monocultures provide higher pollinator and biodiversity support. Covers that include shrubs and other structural diversity that animals use as cover while navigating through urban areas can minimize landscape fragmentation.[134] Lawn alternatives also often require fewer inputs of fertilizer and pesticides to maintain, and less mowing, both of which benefit insect diversity.[41,135]

Human Health Benefits: Reducing the area of lawn monoculture mitigates the harmful effects associated with their upkeep. A high proportion of grass cover in greenspace is associated with poor health[136] and higher heat than diversified vegetation.[137] Traditional lawn monoculture maintenance requires high levels of chemical inputs (pesticides, herbicides, and fertilizers) that have been linked to negative health outcomes including increased risk of cancer,[103] birth defects,[112] and impaired development in children.[138] Lawn chemicals can be deposited on clothing and brought into indoor environments where they persist.[139] These chemicals can be especially hazardous on residential lawns, where they are in close proximity to the home environment. Finally, gas-powered mowers impact air quality.

KEY TENSIONS

▶ Lawns may be optimal for recreational areas, which support sports and other active recreation with strong health benefits. While total elimination of monoculture lawn grass may not be feasible or warranted, the following implementation guidelines can be used to limit the negative effects.

IMPLEMENTATION GUIDANCE

▶ **Use alternative ground cover.** While some areas such as sports fields and gathering locations benefit from traditional turfgrass lawns, large portions of urban greenspace such as traffic medians, pathways, picnic areas, and low traffic areas could benefit from alternatives, including low-height vegetative cover, gravel, dirt, sand, or mulch. These alternatives require fewer harmful inputs, and can provide habitat for insects. See Wildlife-Friendly Management (pg. 66).

▶ **Add structure.** There are many lawns that can be converted to more complex and varied vegetation types that require less intensive irrigation and maintenance. Residential lawns that mainly serve as a visual amenity can be converted into native pollinator gardens (*see image 5*), roadway medians can be planted with drought- and pollution-tolerant shrubs and grasses, and low traffic areas can be converted into structurally complex native landscaping. See Habitat Complexity (pg. 54).

▶ **Plant native grass lawns.** In some bioregions, native grass mixes may provide alternative ground cover to replace traditional turfgrass lawns (*see image 4*). Diverse native grass mixes are slower growing and have 50% lower weed density than traditional lawn monocultures, requiring less mowing and herbicide treatment to maintain.[140] Native grasses have the same tolerance to high traffic as traditional turf.[140]

▶ **Incorporate natural surfaces.** Permeable alternatives to lawns and pavement such as mulch, wood chips, decomposed granite (*see image 3*), and moss can be used for high traffic areas such as pathways and areas around benches and tables.

▶ **Limit maintenance.** See Wildlife-Friendly Management (pg. 66).

1 Permeable pathway in Sunnyvale, CA (SFEI)
2 Permeable path and converted lawn in Sunnyvale, CA (SFEI) **3** Permeable decomposed granite pathway in Sunnyvale, CA (SFEI) **4** Native grass lawn alternative in Alamo Square Park, San Francisco, CA (SFEI) **5** Native front garden converted from lawn in Novato, CA (SFEI)

REFERENCES

1. Jokimäki, J. Occurrence of breeding bird species in urban parks: Effects of park structure and broad-scale variables. *Urban Ecosyst.* **3**, 21–34 (1999).

2. Belaire, J. A., Whelan, C. J. & Minor, E. S. Having our yards and sharing them too: the collective effects of yards on native bird species in an urban landscape. *Ecol. Appl.* **24**, 2132–2143 (2014).

3. Nielsen, A. B., van den Bosch, M., Maruthaveeran, S. & van den Bosch, C. K. Species richness in urban parks and its drivers: A review of empirical evidence. *Urban Ecosyst.* **17**, 305–327 (2014).

4. Tews, J. *et al.* Animal species diversity driven by habitat heterogeneity/diversity: the importance of keystone structures: Animal species diversity driven by habitat heterogeneity. *J. Biogeogr.* **31**, 79–92 (2004).

5. Fuller, R. A., Irvine, K. N., Devine-Wright, P., Warren, P. H. & Gaston, K. J. Psychological benefits of greenspace increase with biodiversity. *Biol. Lett.* **3**, 390–394 (2007).

6. Chawla, L. Benefits of nature contact for children. *J. Plan. Lit.* **30**, 433–452 (2015).

7. Cameron, R. W. *et al.* Where the wild things are! Do urban green spaces with greater avian biodiversity promote more positive emotions in humans? *Urban Ecosyst.* **23**, 301–317 (2020).

8. Dennis, M., Cook, P. A., James, P., Wheater, C. P. & Lindley, S. J. Relationships between health outcomes in older populations and urban green infrastructure size, quality and proximity. *BMC Public Health* **20**, 1–15 (2020).

9. Civitello, D. J. *et al.* Biodiversity inhibits parasites: Broad evidence for the dilution effect. *Proc. Natl. Acad. Sci.* **112**, 8667–8671 (2015).

10. Donovan, G. H., Gatziolis, D., Longley, I. & Douwes, J. Vegetation diversity protects against childhood asthma: results from a large New Zealand birth cohort. *Nat. Plants* **4**, 358–364 (2018).

11. Kim, J.-H., Lee, C. & Sohn, W. Urban Natural Environments, Obesity, and Health-Related Quality of Life among Hispanic Children Living in Inner-City Neighborhoods. *Int. J. Environ. Res. Public. Health* **13**, 121 (2016).

12. Gobster, P. H., Nassauer, J. I., Daniel, T. C. & Fry, G. The shared landscape: what does aesthetics have to do with ecology? *Landsc. Ecol.* **22**, 959–972 (2007).

13. Veen, E. J., Ekkel, E. D., Hansma, M. R. & de Vrieze, A. G. Designing urban green space (Ugs) to enhance health: A methodology. *Int. J. Environ. Res. Public. Health* **17**, 5205 (2020).

14. Faeth, S. H., Bang, C. & Saari, S. Urban biodiversity: patterns and mechanisms. *Ann. N. Y. Acad. Sci.* **1223**, 69–81 (2011).

15. Meyer-Grandbastien, A., Burel, F., Hellier, E. & Bergerot, B. A step towards understanding the relationship between species diversity and psychological restoration of visitors in urban green spaces using landscape heterogeneity. *Landsc. Urban Plan.* **195**, 103728 (2020).

16. Löfvenhaft, K., Björn, C. & Ihse, M. Biotope patterns in urban areas: a conceptual model integrating biodiversity issues in spatial planning. *Landsc. Urban Plan.* **58**, 223–240 (2002).

17. Beller, E. *et al. Historical ecology of the lower Santa Clara River, Ventura River, and Oxnard Plain: an analysis of terrestrial, riverine, and coastal habitats.* https://www.sfei.org/sites/default/files/biblio_files/VenturaCounty_HistoricalEcologyStudy_SFEI_2011_lowres.pdf (2011).

18. Goddard, M. A., Ikin, K. & Lerman, S. B. Ecological and Social Factors Determining the Diversity of Birds in Residential Yards and Gardens. in *Ecology and Conservation of Birds in Urban Environments* (eds. Murgui, E. & Hedblom, M.) 371–397 (Springer International Publishing, 2017). doi:10.1007/978-3-319-43314-1_18.

19. Le Roux, D. S. *et al.* Reduced availability of habitat structures in urban landscapes: implications for policy and practice. *Landsc. Urban Plan.* **125**, 57–64 (2014).

20. Beninde, J., Veith, M. & Hochkirch, A. Biodiversity in cities needs space: a meta-analysis of factors determining intra-urban biodiversity variation. *Ecol. Lett.* **18**, 581–592 (2015).

21. Schebella, M. F., Weber, D., Schultz, L. & Weinstein, P. The wellbeing benefits associated with perceived and measured biodiversity in Australian urban green spaces. *Sustainability* **11**, 802 (2019).

22. Moore, R. C. & Cooper, A. *Nature Play & Learning Places.* http://outdoorplaybook.ca/wp-content/uploads/2015/09/Nature-Play-Learning-Places_v1.5_Jan16.pdf (2014).

23. Bjerke, T., Østdahl, T., Thrane, C. & Strumse, E. Vegetation density of urban parks and perceived appropriateness for recreation. *Urban For. Urban Green.* **5**, 35–44 (2006).

24. Bentrup, G. Conservation Buffers—Design guidelines for buffers, corridors, and greenways. *Gen Tech Rep SRS–109 Asheville NC US Dep. Agric. For. Serv. South. Res. Stn. 110 P* **109**, (2008).

25. Nassauer, J. I. Messy ecosystems, orderly frames. *Landsc. J.* **14**, 161–170 (1995).

26. Threlfall, C. G. *et al.* Increasing biodiversity in urban green spaces through simple vegetation interventions. *J. Appl. Ecol.* **54**, 1874–1883 (2017).

27. Liddicoat, C. *et al.* Landscape biodiversity correlates with respiratory health in Australia. *J. Environ. Manage.* **206**, 113–122 (2018).

28. Spotswood, E. *et al. Re-Oaking Silicon Valley: Building Vibrant Cities with Nature.* https://www.sfei.org/sites/default/files/biblio_files/Re-Oaking%20Silicon%20Valley%20SFEI%20August%202017%20med%20res_B.pdf (2017).

29. Dorner, J. *An introduction to using native plants in restoration projects.* https://www.fs.usda.gov/wildflowers/Native_Plant_Materials/documents/intronatplant.pdf (2002).

30. Berthon, K., Thomas, F. & Bekessy, S. The role of 'nativeness' in urban greening to support animal biodiversity. *Landsc. Urban Plan.* **205**, 103959 (2021).

31. Ehrlich, P. R. & Raven, P. H. Butterflies and Plants: A Study in Coevolution. *Evolution* **18**, 586–608 (1964).

32. Greco, S. E. & Airola, D. A. The importance of native valley oaks (Quercus lobata) as stopover habitat for migratory songbirds in urban Sacramento, California, USA. *Urban For. Urban Green.* **29**, 303–311 (2018).

33. White, J. G., Antos, M. J., Fitzsimons, J. A. & Palmer, G. C. Non-uniform bird assemblages in urban environments: the influence of streetscape vegetation. *Landsc. Urban Plan.* **71**, 123–135 (2005).

34. Prendergast, K. S., Tomlinson, S., Dixon, K. W., Bateman, P. W. & Menz, M. H. Urban native vegetation remnants support more diverse native bee communities than residential gardens in Australia's southwest biodiversity hotspot. *Biol. Conserv.* **265**, 109408 (2022).

35. Pawelek, J. C., Frankie, G. W., Thorp, R. W. & Przybylski, M. Modification of a community garden to attract native bee pollinators in urban San Luis Obispo, California. (2009).

36. Burghardt, K. T., Tallamy, D. W. & Gregory Shriver, W. Impact of Native Plants on Bird and Butterfly Biodiversity in Suburban Landscapes. *Conserv. Biol.* **23**, 219–224 (2009).

37. Pardee, G. L. & Philpott, S. M. Native plants are the bee's knees: local and landscape predictors of bee richness and abundance in backyard gardens. *Urban Ecosyst.* **17**, 641–659 (2014).

38. Hanski, I. *et al.* Environmental biodiversity, human microbiota, and allergy are interrelated. *Proc. Natl. Acad. Sci.* **109**, 8334–8339 (2012).

39. Forristal, L. J., Lehto, X. Y. & Lee, G. The contribution of native species to sense of place. *Curr. Issues Tour.* **17**, 414–433 (2014).

40. Beumer, C. & Martens, P. BIMBY's first steps: a pilot study on the contribution of residential front-yards in Phoenix and Maastricht to biodiversity, ecosystem services and urban sustainability. *Urban Ecosyst.* **19**, 45–76 (2016).

41. Aronson, M. F. *et al.* Biodiversity in the city: key challenges for urban green space management. *Front. Ecol. Environ.* **15**, 189–196 (2017).

42. Grossinger, R. M., Striplen, C. J., Askevold, R. A., Brewster, E. & Beller, E. E. Historical landscape ecology of an urbanized California valley: Wetlands and woodlands in the Santa Clara Valley. *Landsc. Ecol.* **22**, 103–120 (2007).

43. McKay, J. K., Christian, C. E., Harrison, S. & Rice, K. J. "How Local Is Local?"—A Review of Practical and Conceptual Issues in the Genetics of Restoration. *Restor. Ecol.* **13**, 432–440 (2005).

44. Stagoll, K., Lindenmayer, D. B., Knight, E., Fischer, J. & Manning, A. D. Large trees are keystone structures in urban parks. *Conserv. Lett.* **5**, 115–122 (2012).

45. Segar, J. *et al.* Urban conservation gardening in the decade of restoration. *Nat. Sustain.* **5**, 649–656 (2022).

46. Van der Sluijs, J. P. *et al.* Neonicotinoids, bee disorders and the sustainability of pollinator services. *Curr. Opin. Environ. Sustain.* **5**, 293–305 (2013).

47. Cimino, A. M., Boyles, A. L., Thayer, K. A. & Perry, M. J. Effects of neonicotinoid pesticide exposure on human health: a systematic review. *Environ. Health Perspect.* **125**, 155–162 (2017).

48. Wolf, K. L. *et al.* Urban Trees and Human Health: A Scoping Review. *Int. J. Environ. Res. Public. Health* **17**, 4371 (2020).

49. Kim, J.-H., Lee, C., Olvera, N. E. & Ellis, C. D. The Role of Landscape Spatial Patterns on Obesity in Hispanic Children Residing in Inner-City Neighborhoods. *J. Phys. Act. Health* **11**, 1449–1457 (2014).

50. Tsai, W.-L., Floyd, M. F., Leung, Y.-F., McHale, M. R. & Reich, B. J. Urban Vegetative Cover Fragmentation in the U.S. *Am. J. Prev. Med.* **50**, 509–517 (2016).

51. Ziter, C. D., Pedersen, E. J., Kucharik, C. J. & Turner, M. G. Scale-dependent interactions between tree canopy cover and impervious surfaces reduce daytime urban heat during summer. *Proc. Natl. Acad. Sci.* **116**, 7575–7580 (2019).

52. Lovasi, G. S. *et al.* Neighborhood safety and green space as predictors of obesity among preschool children from low-income families in New York City. *Prev. Med.* **57**, 189–193 (2013).

53. Andrusaityte, S. *et al.* Associations between neighbourhood greenness and asthma in preschool children in Kaunas, Lithuania: a case–control study. *BMJ Open* **6**, e010341 (2016).

54. Liu, J. & Slik, F. Are street trees friendly to biodiversity? *Landsc. Urban Plan.* **218**, 104304 (2022).

55. Sand, E. *et al.* Effects of ground surface permeability on the growth of urban linden trees. *Urban Ecosyst.* **21**, 691–696 (2018).

56. Jim, C. Y. Soil volume restrictions and urban soil design for trees in confined planting sites. *J. Landsc. Archit.* **14**, 84–91 (2019).

57. Morales, D. J. The contribution of trees to residential property value. *J. Arboric.* **6**, 305–308 (1980).

58. Donovan, G. H., Prestemon, J. P., Butry, D. T., Kaminski, A. R. & Monleon, V. J. The politics of urban trees: Tree planting is associated with gentrification in Portland, Oregon. *For. Policy Econ.* **124**, 102387 (2021).

59. Walsh, C. J. *et al.* The urban stream syndrome: current knowledge and the search for a cure. *J. North Am. Benthol. Soc.* **24**, 706–723 (2005).

60. Oertli, B. & Parris, K. M. Review: Toward management of urban ponds for freshwater biodiversity. *Ecosphere* **10**, e02810 (2019).

61. Dickman, C. R. Habitat fragmentation and vertebrate species richness in an urban environment. *J. Appl. Ecol.* 337–351 (1987).

62. Chester, E. T. & Robson, B. J. Anthropogenic refuges for freshwater biodiversity: their ecological characteristics and management. *Biol. Conserv.* **166**, 64–75 (2013).

63. Hassall, C. The ecology and biodiversity of urban ponds. *Wiley Interdiscip. Rev. Water* **1**, 187–206 (2014).

64. Čerba, D. & Hamerlík, L. Fountains—overlooked small water bodies in the urban areas. in *Small Water Bodies of the Western Balkans* 73–91 (Springer, 2022).

65. Tryjanowski, P. *et al.* Summer water sources for temperate birds: use, importance, and threats. *Eur. Zool. J.* **89**, 913–926 (2022).

66. Grilli, G., Mohan, G. & Curtis, J. Public park attributes, park visits, and associated health status. *Landsc. Urban Plan.* **199**, 103814 (2020).

67. Gunawardena, K.-C., Wells, M. & Kershaw, T. Utilising green and bluespace to mitigate urban heat island intensity. *Sci. Total Environ.* **584–585**, 1040–1055 (2017).

68. Schipperijn, J., Bentsen, P., Troelsen, J., Toftager, M. & Stigsdotter, U. K. Associations between physical activity and characteristics of urban green space. *Urban For. Urban Green.* **12**, 109–116 (2013).

69. Völker, S., Matros, J. & Claßen, T. Determining urban open spaces for health-related appropriations: a qualitative analysis on the significance of blue space. *Environ. Earth Sci.* **75**, 1–18 (2016).

70. Huynh, Q., Craig, W., Janssen, I. & Pickett, W. Exposure to public natural space as a protective factor for emotional well-being among young people in Canada. *BMC Public Health* **13**, 1–14 (2013).

71. Sievers, M., Parris, K. M., Swearer, S. E. & Hale, R. Stormwater wetlands can function as ecological traps for urban frogs. *Ecol. Appl.* **28**, 1106–1115 (2018).

72. Schmidt, K. A. & Ostfeld, R. S. Biodiversity and the dilution effect in disease ecology. *Ecology* **82**, 609–619 (2001).

73. Buchholz, T. A., Madary, D. A., Bork, D. & Younos, T. Stream restoration in urban environments: concept, design principles, and case studies of stream daylighting. *Sustain. Water Manag. Urban Environ.* 121–165 (2016).

74. Tresch, S. *et al.* Direct and indirect effects of urban gardening on aboveground and belowground diversity influencing soil multifunctionality. *Sci. Rep.* **9**, 9769 (2019).

75. Barthel, S., Folke, C. & Colding, J. Social–ecological memory in urban gardens—Retaining the capacity for management of ecosystem services. *Glob. Environ. Change* **20**, 255–265 (2010).

76. Sorace, A. Value to wildlife of urban-agricultural parks: a case study from Rome urban area. *Environ. Manage.* **28**, 547–560 (2001).

77. Clucas, B., Parker, I. D. & Feldpausch-Parker, A. M. A systematic review of the relationship between urban agriculture and biodiversity. *Urban Ecosyst.* **21**, 635–643 (2018).

78. Lin, B. B., Philpott, S. M. & Jha, S. The future of urban agriculture and biodiversity-ecosystem services: Challenges and next steps. *Basic Appl. Ecol.* **16**, 189–201 (2015).

79. Rudd, H., Vala, J. & Schaefer, V. Importance of backyard habitat in a comprehensive biodiversity conservation strategy: a connectivity analysis of urban green spaces. *Restor. Ecol.* **10**, 368–375 (2002).

80. Soga, M., Gaston, K. J. & Yamaura, Y. Gardening is beneficial for health: A meta-analysis. *Prev. Med. Rep.* **5**, 92–99 (2017).

81. Thompson, R. Gardening for health: a regular dose of gardening. *Clin. Med.* **18**, 201–205 (2018).

82. Gonzalez, M. T., Hartig, T., Patil, G. G., Martinsen, E. W. & Kirkevold, M. Therapeutic horticulture in clinical depression: a prospective study of active components. *J. Adv. Nurs.* **66**, 2002–2013 (2010).

83. Ghanbari, S., Jafari, F., Bagheri, N., Neamtolahi, S. & Shayanpour, R. Study of the effect of using purposeful activity (gardening) on depression of female resident in Golestan Dormitory of Ahvaz Jundishapur University of Medical Sciences. *J. Rehabil. Sci. Res.* **2**, 8–11 (2015).

84. Wilson, J. F. & Christensen, K. M. The relationship between gardening and depression among individuals with disabilities. *J. Ther. Hortic.* **21**, 28–41 (2011).

85. Van den Berg, A. E., van Winsum-Westra, M., De Vries, S. & Van Dillen, S. M. Allotment gardening and health: a comparative survey among allotment gardeners and their neighbors without an allotment. *Environ. Health* **9**, 1–12 (2010).

86. Ulrich, R. S. Effects of gardens on health outcomes: theory and research. in *Healing Gardens: Therapeutic Benefits and Design Recommendations* (1999).

87. Franco, L. S., Shanahan, D. F. & Fuller, R. A. A Review of the Benefits of Nature Experiences: More Than Meets the Eye. *Int. J. Environ. Res. Public. Health* **14**, 864 (2017).

88. Attanayake, C. P. *et al.* Field evaluations on soil plant transfer of lead from an urban garden soil. *J. Environ. Qual.* **43**, 475–487 (2014).

89. Russo, A., Escobedo, F. J., Cirella, G. T. & Zerbe, S. Edible green infrastructure: An approach and review of provisioning ecosystem services and disservices in urban environments. *Agric. Ecosyst. Environ.* **242**, 53–66 (2017).

90. Smith, R. M., Thompson, K., Hodgson, J. G., Warren, P. H. & Gaston, K. J. Urban domestic gardens (IX): Composition and richness of the vascular plant flora, and implications for native biodiversity. *Biol. Conserv.* **129**, 312–322 (2006).

91. Turner-Skoff, J. B. & Cavender, N. The benefits of trees for livable and sustainable communities. *Plants People Planet* **1**, 323–335 (2019).

92. Briffett, C. Is managed recreational use compatible with effective habitat and wildlife occurrence in urban open space corridor systems? *Landsc. Res.* **26**, 137–163 (2001).

93. Russell, W., Shulzitski, J. & Setty, A. Evaluating wildlife response to coastal dune habitat restoration in San Francisco, California. *Ecol. Restor.* **27**, 439–448 (2009).

94. Jachowski, D. S., Slotow, R. & Millspaugh, J. J. Good virtual fences make good neighbors: opportunities for conservation. *Anim. Conserv.* **17**, 187–196 (2014).

95. Mateer, T. J. Developing connectedness to nature in urban outdoor settings: a potential pathway through awe, solitude, and leisure. *Front. Psychol.* 4137 (2022).

96. Abrams, K. M., Molder, A. L., Nankey, P. & Leong, K. Encouraging Respectful Wildlife Viewing Among Tourists: Roles for Social Marketing, Regulatory Information, Symbolic Barriers, and Enforcement. *Soc. Mark. Q.* 15245004231153084 (2023).

97. Konrad, L. & Levine, A. Controversy over beach access restrictions at an urban coastal seal rookery: Exploring the drivers of conflict escalation and endurance at Children's Pool Beach in La Jolla, CA. *Mar. Policy* **132**, 104659 (2021).

98. NOAA. California Balances Habitat Conservation with Public Beach Access. https://coast.noaa.gov/states/stories/balance-habitat-conservation-with-public-beach-access.html (2020).

99. Kretser, H. E. *et al.* Management Recommendations for Balancing Public Access and Species Conservation on Protected Lands. (2022).

100. 2M Associates, PlaceWorks, Questa Engineering Corporation & DeRobertis, M. *San Francisco Bay Trail Design Guidelines and Toolkit.* https://mtc.ca.gov/sites/default/files/documents/2021-10/Bay-Trail-Design-Guidelines-and-Toolkit.pdf (2016).

101. Stigsdotter, U. & Peschardt, K. Evidence for Designing Health Promoting Pocket Parks. *Int. J. Archit. Res.* **8**, 149–164 (2014).

102. Sehrt, M., Bossdorf, O., Freitag, M. & Bucharova, A. Less is more! Rapid increase in plant species richness after reduced mowing in urban grasslands. *Basic Appl. Ecol.* **42**, 47–53 (2020).

103. Sabarwal, A., Kumar, K. & Singh, R. P. Hazardous effects of chemical pesticides on human health–Cancer and other associated disorders. *Environ. Toxicol. Pharmacol.* **63**, 103–114 (2018).

104. Banks, J. L. & McConnell, R. National emissions from lawn and garden equipment. in *International Emissions Inventory Conference, San Diego, April* vol. 16 (2015).

105. Fischer, L. K. *et al.* Public attitudes toward biodiversity-friendly greenspace management in Europe. *Conserv. Lett.* **13**, e12718 (2020).

106. Kane, B., Warren, P. S. & Lerman, S. B. A broad scale analysis of tree risk, mitigation and potential habitat for cavity-nesting birds. *Urban For. Urban Green.* **14**, 1137–1146 (2015).

107. Ehler, L. E. Integrated pest management (IPM): definition, historical development and implementation, and the other IPM. *Pest Manag. Sci.* **62**, 787–789 (2006).

108. Spotswood, E. *et al.* Making Nature's City: A Science-Based Framework for Building Urban Biodiversity. *San Franc. Estuary Inst. Publ.* (2019).

109. Watson, C. J., Carignan-Guillemette, L., Turcotte, C., Maire, V. & Proulx, R. Ecological and economic benefits of low-intensity urban lawn management. *J. Appl. Ecol.* **57**, 436–446 (2020).

110. Wintergerst, J., Kästner, T., Bartel, M., Schmidt, C. & Nuss, M. Partial mowing of urban lawns supports higher abundances and diversities of insects. *J. Insect Conserv.* **25**, 797–808 (2021).

111. Van Bruggen, A. H. *et al.* Environmental and health effects of the herbicide glyphosate. *Sci. Total Environ.* **616**, 255–268 (2018).

112. Sapbamrer, R. & Hongsibsong, S. Effects of prenatal and postnatal exposure to organophosphate pesticides on child neurodevelopment in different age groups: a systematic review. *Environ. Sci. Pollut. Res.* **26**, 18267–18290 (2019).

113. Teysseire, R. *et al.* Assessment of residential exposures to agricultural pesticides: A scoping review. *PloS One* **15**, e0232258 (2020).

114. McLaughlin, A. & Mineau, P. The impact of agricultural practices on biodiversity. *Agric. Ecosyst. Environ.* **55**, 201–212 (1995).

115. Geiger, F. *et al.* Persistent negative effects of pesticides on biodiversity and biological control potential on European farmland. *Basic Appl. Ecol.* **11**, 97–105 (2010).

116. Quistberg, R. D., Bichier, P. & Philpott, S. M. Landscape and local correlates of bee abundance and species richness in urban gardens. *Environ. Entomol.* **45**, 592–601 (2016).

117. Davies, T. W. & Smyth, T. Why artificial light at night should be a focus for global change research in the 21st century. *Glob. Change Biol.* **24**, 872–882 (2018).

118. Barghini, A. & de Medeiros, B. A. Artificial lighting as a vector attractant and cause of disease diffusion. *Environ. Health Perspect.* **118**, 1503–1506 (2010).

119. Cho, Y. *et al.* Effects of artificial light at night on human health: A literature review of observational and experimental studies applied to exposure assessment. *Chronobiol. Int.* **32**, 1294–1310 (2015).

120. Ramsay, M. & Newton, R. The effect of better street lighting on crime and fear: A review. (1991).

121. Boomsma, C. & Steg, L. Feeling Safe in the Dark: Examining the Effect of Entrapment, Lighting Levels, and Gender on Feelings of Safety and Lighting Policy. *Environ. Behav.* **46**, (2012).

122. Rahm, J., Sternudd, C. & Johansson, M. "In the evening, I don't walk in the park": The interplay between street lighting and greenery in perceived safety. *Urban Des. Int.* **26**, 42–52 (2021).

123. Cohen, D. A. *et al.* Public parks and physical activity among adolescent girls. *Pediatrics* **118**, e1381–e1389 (2006).

124. Rakonjac, I. *et al.* Increasing the Livability of Open Public Spaces during Nighttime: The Importance of Lighting in Waterfront Areas. *Sustainability* **14**, 6058 (2022).

125. Gaston, K. J., Davies, T. W., Bennie, J. & Hopkins, J. Reducing the ecological consequences of night-time light pollution: options and developments. *J. Appl. Ecol.* **49**, 1256–1266 (2012).

126. Dominoni, D. M., Quetting, M. & Partecke, J. Long-term effects of chronic light pollution on seasonal functions of European blackbirds (Turdus merula). *PLoS One* **8**, e85069 (2013).

127. Jägerbrand, A. K. & Bouroussis, C. A. Ecological impact of artificial light at night: effective strategies and measures to deal with protected species and habitats. *Sustainability* **13**, 5991 (2021).

128. Davies, T. W., Bennie, J., Inger, R., De Ibarra, N. H. & Gaston, K. J. Artificial light pollution: are shifting spectral signatures changing the balance of species interactions? *Glob. Change Biol.* **19**, 1417–1423 (2013).

129. Longcore, T. *et al.* Rapid assessment of lamp spectrum to quantify ecological effects of light at night. *J. Exp. Zool. Part Ecol. Integr. Physiol.* **329**, 511–521 (2018).

130. Deichmann, J. L. *et al.* Reducing the blue spectrum of artificial light at night minimises insect attraction in a tropical lowland forest. *Insect Conserv. Divers.* **14**, 247–259 (2021).

131. Witherington, B. E. Behavioral responses of nesting sea turtles to artificial lighting. *Herpetologica* 31–39 (1992).

132. Rebke, M. *et al.* Attraction of nocturnally migrating birds to artificial light: The influence of colour, intensity and blinking mode under different cloud cover conditions. *Biol. Conserv.* **233**, 220–227 (2019).

133. International Dark Sky Association & Illuminating Engineering Society of North America. *Model Lighting Ordinance.* https://www.usgbc.org/sites/default/files/mlo_final_june2011.pdf (2011).

134. Bennett, A. F., Henein, K. & Merriam, G. Corridor use and the elements of corridor quality: Chipmunks and fencerows in a farmland mosaic. *Biol. Conserv.* **68**, 155–165 (1994).

135. Brede, D. *Turfgrass maintenance reduction handbook: Sports, lawns, and golf.* (John Wiley & Sons, 2000).

136. Mears, M., Brindley, P., Jorgensen, A., Ersoy, E. & Maheswaran, R. Greenspace spatial characteristics and human health in an urban environment: An epidemiological study using landscape metrics in Sheffield, UK. *Ecol. Indic.* **106**, 105464 (2019).

137. Francoeur, X. W., Dagenais, D., Paquette, A., Dupras, J. & Messier, C. Complexifying the urban lawn improves heat mitigation and arthropod biodiversity. *Urban For. Urban Green.* **60**, 127007 (2021).

138. Pascale, A. & Laborde, A. Impact of pesticide exposure in childhood. *Rev. Environ. Health* **35**, 221–227 (2020).

139. Robbins, P. & Birkenholtz, T. Turfgrass revolution: measuring the expansion of the American lawn. *Land Use Policy* **20**, 181–194 (2003).

140. Simmons, M., Bertelsen, M., Windhager, S. & Zafian, H. The performance of native and non-native turfgrass monocultures and native turfgrass polycultures: An ecological approach to sustainable lawns. *Ecol. Eng.* **37**, 1095–1103 (2011).

◀ Swallowtail butterfly on sunflowers
(Brynn Pedrick, Unsplash)

www.ingramcontent.com/pod-product-compliance
Lightning Source LLC
Chambersburg PA
CBRC091141030426
42335CB00010B/209

9 781950 313136